W0037105

**Lectures in Mathematics**
**ETH Zürich**
Department of Mathematics
Research Institute of Mathematics

Managing Editor:
Michael Struwe

Wolfgang Bangerth
Rolf Rannacher
**Adaptive Finite
Element Methods
for Differential Equations**

Springer Basel AG

Authors' addresses:

Wolfgang Bangerth
TICAM
The University of Texas at Austin
201 E. 24th Street
Austin, TX 78712
USA
e-mail: bangerth@ticam.utexas.edu

Rolf Rannacher
Institute of Applied Mathematics
University of Heidelberg
Im Neuenheimer Feld 293
69120 Heidelberg
Germany
e-mail: rannacher@iwr.uni-heidelberg.de

2000 Mathematical Subject Classification 65L60, 65L70, 65M60, 65Nxx, 74S05, 76M10

A CIP catalogue record for this book is available from the
Library of Congress, Washington D.C., USA


Bibliographic information published by Die Deutsche Bibliothek
Die Deutsche Bibliothek lists this publication in the Deutsche Nationalbibliografie; detailed
bibliographic data is available in the Internet at <http://dnb.ddb.de>.

ISBN 978-3-7643-7009-1      ISBN 978-3-0348-7605-6 (eBook)
DOI 10.1007/978-3-0348-7605-6

This work is subject to copyright. All rights are reserved, whether the whole or part of the material
is concerned, specifically the rights of translation, reprinting, re-use of illustrations, recitation,
broadcasting, reproduction on microfilms or in other ways, and storage in data banks. For any kind
of use permission of the copyright owner must be obtained.

© 2003 Springer Basel AG
Originally published by Birkhäuser Verlag, Basel - Boston - Berlin in 2003
Printed on acid-free paper produced from chlorine-free pulp. TCF ∞

9 8 7 6 5 4 3 2 1                                                    www.birkhauser-science.com

# Contents

# Preface

These Lecture Notes have been compiled from the material presented by the second author in a lecture series ('Nachdiplomvorlesung') at the Department of Mathematics of the ETH Zürich during the summer term 2002. Concepts of 'self-adaptivity' in the numerical solution of differential equations are discussed with emphasis on Galerkin finite element methods. The key issues are *a posteriori* error estimation and *automatic* mesh adaptation. Besides the traditional approach of energy-norm error control, a new duality-based technique, the *Dual Weighted Residual* method (or shortly *DWR* method) for goal-oriented error estimation is discussed in detail. This method aims at economical computation of arbitrary quantities of physical interest by properly adapting the computational mesh. This is typically required in the design cycles of technical applications. For example, the drag coefficient of a body immersed in a viscous flow is computed, then it is minimized by varying certain control parameters, and finally the stability of the resulting flow is investigated by solving an eigenvalue problem. 'Goal-oriented' adaptivity is designed to achieve these tasks with minimal cost.

The basics of the DWR method and various of its applications are described in the following survey articles:

R. Rannacher [114], *Error control in finite element computations*. In: Proc. of Summer School Error Control and Adaptivity in Scientific Computing (H. Bulgak and C. Zenger, eds), pp. 247–278. Kluwer Academic Publishers, 1998.

M. Braack and R. Rannacher [42], *Adaptive finite element methods for low-Mach-number flows with chemical reactions*. In: 30th Computational Fluid Dynamics (H. Deconinck, ed.), Vol. 1999-03 of Lecture Series, von Karman Institute for Fluid Dynamics, Brussels, 1999.

R. Becker and R. Rannacher [31], *An optimal control approach to error estimation and mesh adaptation in finite element methods*. In: Acta Numerica 2000 (A. Iserles, ed.), pp. 1–101, Cambridge University Press, 2001.

R. Rannacher [117], *Duality techniques for error estimation and mesh adaptation in finite element methods.* In: Adaptive Finite Elements in Linear and Nonlinear Solid and Structural Mechanics (E. Stein, ed.), Vol. 416 of CISM Courses and Lectures, Springer, 2002, to appear.

R. Rannacher and F.-T. Suttmeier [121]. *Error estimation and adaptive mesh design for FE models in elasto-plasticity.* In: Error-Controlled Adaptive FEMs in Solid Mechanics (E. Stein, ed.), John Wiley, 2002, to appear.

Much of the contents of these Lecture Notes is taken from the above articles but new material has also been added. At the end of each chapter some exercises are posed in order to assist the interested reader in better understanding the presented concepts. Solutions and accompanying remarks are given in the Appendix. For these practical exercises, sample programs are provided at:
http://gaia.iwr.uni-heidelberg.de/httpdoc/Research/software.AFEMforDE.html.

# Chapter 1

# Introduction

We begin with a brief introduction to the philosophy underlying the approach to self-adaptivity which will be discussed in these Lecture Notes. Let the goal of a simulation be the accurate and efficient computation of the value of a functional $J(u)$, the 'target quantity', with accuracy $TOL$ from the solution $u$ of a continuous model by using an approximative discrete model of dimension $N$:

$$\mathcal{A}(u) = 0, \qquad \mathcal{A}_h(u_h) = 0.$$

The evaluation of the solution by the functional $J(\cdot)$ represents what exactly we want to know of a solution. Then, the goal of adaptivity is the 'optimal' use of computing resources according to either one of the following principles:

- Minimal work $N$ for prescribed accuracy $TOL$,

$$N \to \min, \quad TOL \text{ given}.$$

- Maximal accuracy for prescribed work,

$$TOL \to \min, \quad N \text{ given}.$$

These goals are traditionally approached by automatic mesh adaptation on the basis of local 'error indicators' taken from the computed solution, assuming that this can indicate local roughness of the 'continuous' solution. The main ingredients of this process are:

- rigorous a posteriori error estimates in terms of data and the computed solution employing information about the continuous problem;

- local error indicators extracted from the a posteriori error estimates;

- automatic mesh adaptation according to certain refinement strategies based on the local error indicators.

We will demonstrate by examples that the appropriate choice of each of these steps is crucial for an economical simulation. Inappropriate realizations which violate the characteristic features of the underlying problem may drastically reduce the efficiency and accuracy.

The traditional approach to adaptivity aims at estimating the error with respect to the generic energy norm of the problem, or the global $L^2$-norm. However this is generally not what applications need. In the following, we will present a collection of examples for such situations, where one is really interested in computing locally defined quantities.

## 1.1   A first example: computation of drag coefficient

In order to illustrate the role of adaptivity in the design of a computational mesh, we consider a viscous incompressible flow around a cylinder in a channel with a narrowed outlet as shown in Figure 1.1. The flow quantities, velocity $v$ and pressure $p$, are determined by the classical 'incompressible' Navier-Stokes equations

$$\partial_t v - \nu \Delta v + v \cdot \nabla v + \nabla p = f, \quad \nabla \cdot v = 0.$$

The configuration is two-dimensional, with Poisseuille inflow, and Reynolds number $\mathrm{Re} = 50$, such that the flow is laminar and stationary. The narrowing of the outlet causes a so-called *corner singularity* of the pressure.

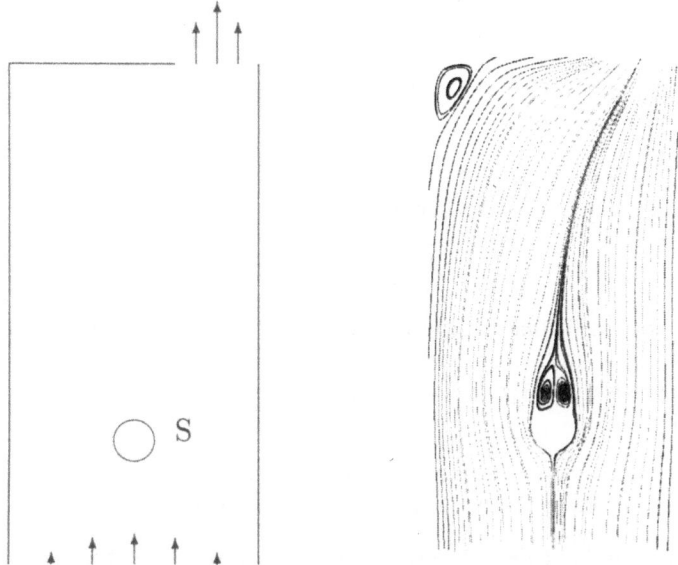

Figure 1.1: *Configuration and streamline plot for flow around a cylinder (* $\mathrm{Re} = 50$ *)*.

The goal is the accurate computation of the corresponding drag coefficient

$$J(v,p) := c_{\text{drag}} := \frac{2}{\bar{U}^2 D} \int_S n^T (2\nu\tau - pI) d \, ds,$$

of the obstacle, where $S$ is the surface of the body, $D$ its diameter, $\bar{U}$ the maximal inflow velocity, $\tau := \frac{1}{2}(\nabla v + \nabla v^T)$ the strain tensor, and $d = (0,1)^T$ the main flow direction.

In order to control the mesh adaptation process in this simulation, one may find good reasons to use either of the following heuristic refinement indicators $\eta_K$ on the mesh cells $K$:

- *Vorticity:*   $\eta_K := h_K \|\nabla \times v_h\|_K.$

- *First-order pressure gradient:*   $\eta_K := h_K \|\nabla p_h\|_K.$

- *Second-order velocity gradient:*   $\eta_K := h_K \|\nabla_h^2 v_h\|_K.$

- *Residual-based indicator:*   $\eta_K := h_K \|R_h\|_K + h_K^{1/2} \|r_h\|_{\partial K} + h_K \|\nabla \cdot v_h\|_K,$

$$R_{h|K} := f + \nu \Delta v_h - v_h \cdot \nabla v_h - \nabla p_h,$$

$$r_{h|\Gamma} := \left\{ \begin{array}{ll} \frac{1}{2}[\nu \partial_n v_h - n p_h], & \text{if } \Gamma \not\subset \partial\Omega \\ 0, & \text{if } \Gamma \subset \Gamma_{\text{rigid}} \cup \Gamma_{\text{in}} \\ -\nu \partial_n v_h + n p_h, & \text{if } \Gamma \subset \Gamma_{\text{out}} \end{array} \right\},$$

where $n$ is the normal unit vector and $[\cdot]$ the jump across cell interfaces.

The vorticity as well as the pressure and velocity gradient indicators measure the 'smoothness' of $\{v_h, p_h\}$, while the heuristical residual-based indicator additionally contains information about local conservation of momentum and mass. As competitors, we additionally consider global uniform refinement and refinement using a new approach, called *DWR method* (**D**ual **W**eighted **R**esidual method) which uses the same residual terms as in the heuristic residual-based indicators but multiplied by weights obtained by solving a global 'dual' problem. This method will be systematically developed in these Lecture Notes.

The above test case has been designed in order to demonstrate the ability of the different refinement indicators to produce meshes on which the main features of the flow, such as boundary layers along rigid walls, vortices behind the cylinder, and the corner singularity at the outlet, are sufficiently resolved. Figure 1.2 shows locally adapted meshes obtained on the basis of the different refinement indicators. The results shown in Figure 1.3 demonstrate that in this case the two ad hoc indicators involving only the norm of vorticity or pressure gradient are even less efficient than simple uniform refinement. This demonstrates that a systematic approach to goal-oriented mesh adaptation is needed which not only takes into account local properties of the solution but also the global dependence of the error in the target quantity on these properties.

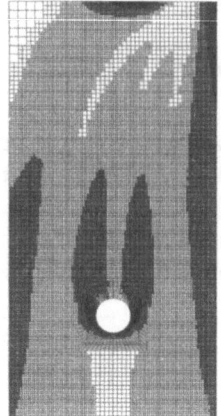

Figure 1.2: *Meshes with 5,000 cells obtained by the vorticity indicator (left), the heuristic residual-based indicator (middle), and the new DWR indicator (right).*

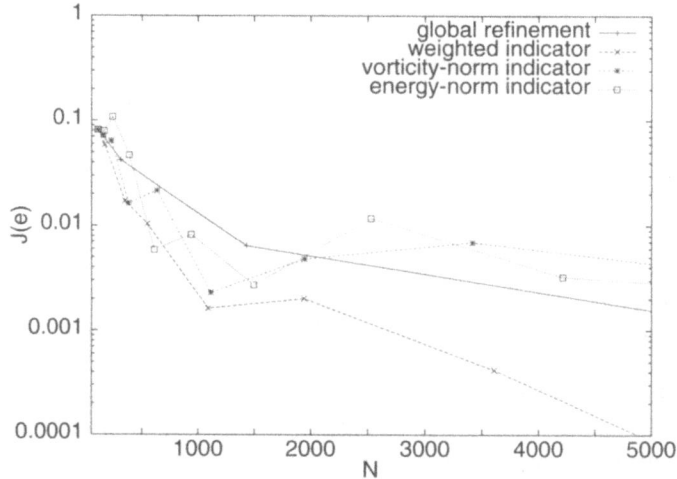

Figure 1.3: *Error $J(e)$ in the drag coefficient versus number of cells $N$, for uniform refinement, the weighted indicator obtained by the DWR approach, the vorticity indicator, and the heuristic residual-based indicator.*

## 1.2   The need for 'goal-oriented' mesh adaptation

Let us illustrate the need for 'goal-oriented' mesh adaptation by two further examples of different types of complexity: a 3-d flow problem where only on locally adapted meshes sufficient accuracy is achieved with acceptable costs, and a 2-d flow problem in which the complication arises through the strong interaction of mass transport and heat transfer. In both situations, the main problem is the generation of error estimates which reflect the local and global dependency of the error on the locally observed solution properties. In fact, usually mesh adaptation in solving a coupled system of equations for a set of physical quantities $u = (u_h^i)_{i=1}^n$ is based on 'smoothness' or 'residual' information like

$$\eta_K := \sum_{i=1}^n \omega_K^i \|D_h^2 u_h^i\|_K \qquad \text{or} \qquad \eta_K = \sum_{i=1}^n \omega_K^i \|R_i(u_h)\|_K.$$

Here, $D_h^2 u_h^i$ stands for certain second-order difference quotients, and $R_i(u_h)$ are certain residuals as introduced in the previous example. Both kinds of indicators are easily evaluated from the computed solution and are widely used in practice. The proper choice of the weights $\omega_K^i$ is crucial for the effectivity of the adaptation process. They should include both a scaling due to different orders of magnitude of the solution components $u^i$, as well as the influence of the present cell $K$ on the requested quantity of interest.

## A cylinder flow benchmark in 3-D

We consider the 3-dimensional flow in a channel around a cylinder with square cross section as shown in Figure 1.4. The Reynolds number is Re = 20, such that the flow is laminar and stationary. This is part of a benchmark suite for the computation of viscous, incompressible fluid flow (see Schäfer and Turek [124]).

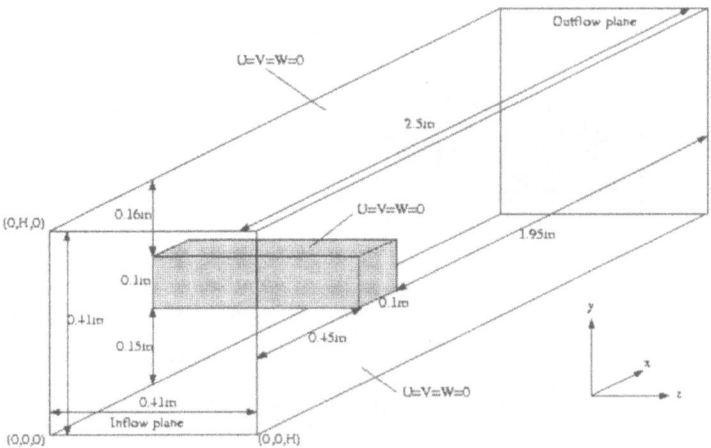

Figure 1.4: *Configuration of 3D flow around a square cylinder in a channel.*

The goal is again the accurate computation of the drag coefficient

$$J(v,p) := c_d = \frac{2}{\bar{U}^2 DH} \int_S n^T (2\nu\tau - pI) d\,\mathrm{ds},$$

where $D$, $H$ are geometrical parameters, $\bar{U}$ the maximum inflow velocity and $d = (1,0,0)^T$ the main flow direction. The desired accuracy is TOL $\sim 1\%$ which turns out to be a rather demanding task even for this simple flow situation.

In Table 1.1, we compare the efficiency of global uniform refinement against that of local refinement on the basis of a heuristic residual-based indicator and the DWR method as already used in the first example. The superiority of mesh adaptation by the DWR method is clearly seen.

Table 1.1: *Drag results: a) $Q_2/Q_1$-element with global refinement, b) $Q_1/Q_1$-element with local refinement by heuristic residual indicator, c) $Q_1/Q_1$-element with local refinement by DWR method; the first mesh level on which an error smaller than 1% is achieved is indicated in boldface; from Braack et al. [40].*

| a) $N$ | $c_d$ | b) $N$ | $c_d$ | c) $N$ | $c_d$ |
|---:|---|---:|---|---:|---|
| $15,960$ | 8.2559 | $3,696$ | 12.7888 | $3,696$ | 12.7888 |
| $117,360$ | 7.9766 | $21,512$ | 8.7117 | $8,456$ | 9.8262 |
| $899,040$ | 7.8644 | $80,864$ | 7.9505 | $15,768$ | 8.1147 |
| $\mathbf{7,035,840}$ | 7.8193 | $182,352$ | 7.9142 | $30,224$ | 8.1848 |
| $55,666,560$ | 7.7959 | $473,000$ | 7.8635 | $\mathbf{84,832}$ | 7.8282 |
| $-$ | $-$ | $\mathbf{1,052,000}$ | 7.7971 | $162,680$ | 7.7788 |
| $\infty$ | 7.7730 | $\infty$ | 7.7730 | $\infty$ | 7.7730 |

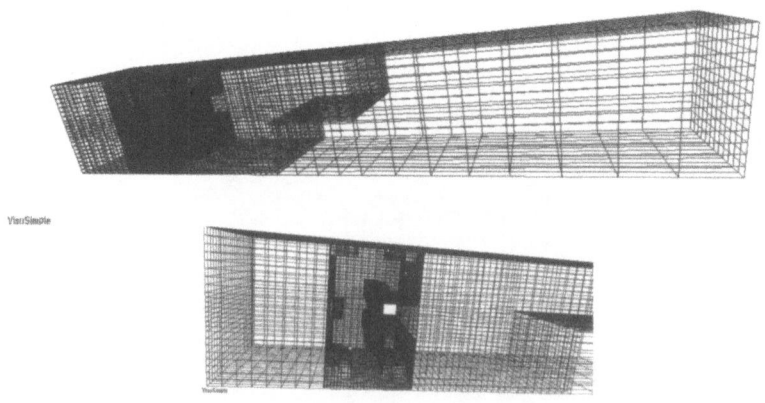

Figure 1.5: *Refined mesh and zoom into the cylinder vicinity obtained by the heuristic residual-based indicator; from Braack et al. [40].*

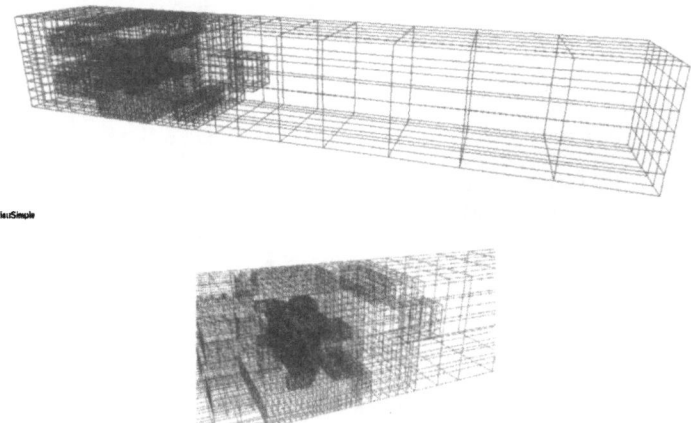

Figure 1.6: *Refined mesh and zoom into the cylinder vicinity obtained by the DWR method; from Braack et al. [40].*

Figures 1.5 and 1.6 show meshes which have been obtained by using refinement on the basis of the heuristic residual-based indicator and by the DWR method. The heuristic indicator fails to properly refine the area behind the cylinder which causes its poorer drag approximation.

## A heat-driven cavity benchmark in 2-D

We consider a 2-dimensional cavity flow. The flow in a square box with side length $L = 1$ (see Figure 1.7) is driven by a temperature difference $\theta_h - \theta_c = 720\,K$, between the left ('hot') and the right ('cold') wall, under the action of gravity $g$ in the vertical direction. The Rayleigh number is $Ra \sim 10^6$ making this problem computationally demanding. Here, the quantity to be computed is the average Nusselt number (mean heat flux) along the cold wall defined by

$$J(u) \;=\; \langle \mathrm{Nu} \rangle_c := \frac{\mathrm{Pr}}{0.3\mu_0\theta_0} \int_{\Gamma_{cold}} \kappa \partial_n \theta \, ds,$$

where $\mathrm{Pr}$ is the Prandtl number and $\mu_0$, $\theta_0$ are certain reference values for viscosity and temperature. The underlying mathematical model is the *low-Mach number approximation* of the stationary compressible Navier-Stokes equations which is expressed in terms of the set of primitive variables $u = \{v, p, \theta\}$ denoting velocity, pressure and temperature. In this case, due to the large temperature difference, the usual Boussinesq approximation is not sufficient (see Becker and Braack [24]).

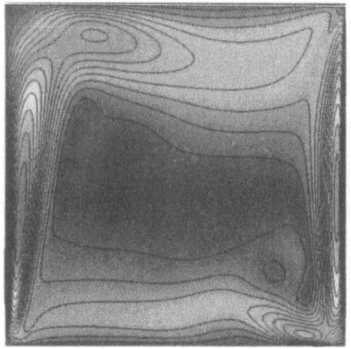
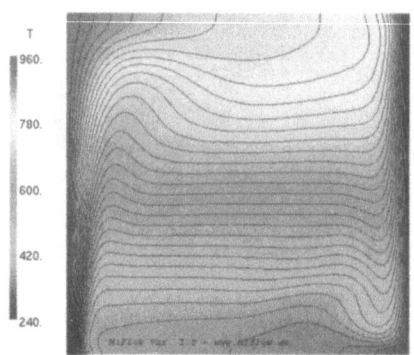

Figure 1.7: *Flow in heat-driven cavity: velocity norm isolines (left) and temperature isolines (right); from Becker and Braack [24].*

The meshes shown in Figure 1.8 indicate that the heuristic residual-based indicator induces mesh refinement mainly in those areas where the velocity is dominant while the weighted error indicator obtained by the DWR method puts more emphasis on the region along the hot boundary where the temperature gradient is dominant. The latter, given the quantity we want to compute, seems to be more important for capturing the heat transfer through the cavity. This is confirmed by the results shown in Table 1.2 which show that on the meshes generated by the properly weighted error indicators the accuracy in computing the Nusselt number is better by almost one order.

Table 1.2: *Computation of the Nusselt number in the heat-driven cavity by the heuristic residual-based indicator (left) and the weighted indicator by the DWR method (right); comparable error magnitudes are indicated in boldface; from Becker and Braack [24].*

| N | $\langle Nu \rangle_c$ | error | N | $\langle Nu \rangle_c$ | error |
|---|---|---|---|---|---|
| 524 | -9.09552 | $4.1 \cdot 10^{-1}$ | 523 | -8.86487 | $1.8 \cdot 10^{-1}$ |
| 945 | -8.67201 | $1.5 \cdot 10^{-2}$ | 945 | -8.71941 | $3.3 \cdot 10^{-2}$ |
| 1708 | -8.49286 | $1.9 \cdot 10^{-1}$ | 1717 | -8.66898 | $1.8 \cdot 10^{-2}$ |
| 3108 | -8.58359 | $1.0 \cdot 10^{-1}$ | 5530 | -8.67477 | $1.2 \cdot 10^{-2}$ |
| 5656 | -8.59982 | $8.7 \cdot 10^{-2}$ | **9728** | -8.68364 | $3.0 \cdot 10^{-3}$ |
| 18204 | -8.64775 | $3.9 \cdot 10^{-2}$ | 17319 | -8.68744 | $8.5 \cdot 10^{-4}$ |
| 32676 | -8.66867 | $1.8 \cdot 10^{-2}$ | 31466 | -8.68653 | $6.9 \cdot 10^{-5}$ |
| 58678 | -8.67791 | $8.7 \cdot 10^{-3}$ | | | |
| **79292** | -8.67922 | $7.4 \cdot 10^{-3}$ | | | |

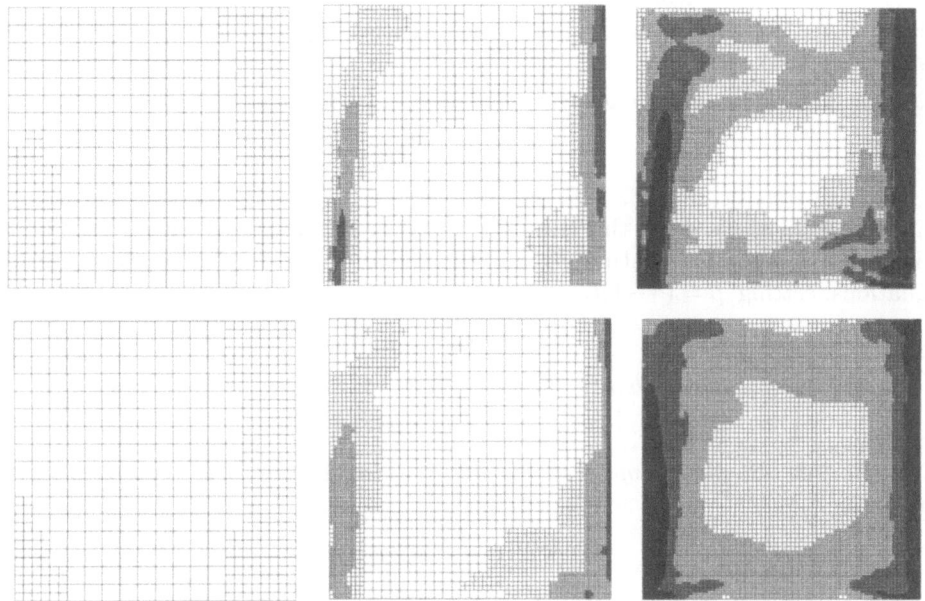

Figure 1.8: *Sequence of refined meshes for the heat-driven cavity with $N \approx$ 500, 5500, 56000 cells: heuristic residual-based indicator (top row), weighted error indicator by the DWR method (bottom row); from Becker and Braack [24].*

## 1.3 Further examples of goal-oriented simulation

In this section, let us present a collection of further examples from different areas in which 'goal-oriented' adaptivity is needed and has already proven to be superior over other more ad-hoc approaches. Some of these examples will be discussed in more detail in the course of these Lecture Notes.

*Example* 1.1. *CARS signal in a flow reactor* (Becker et al. [26]):

$$J(p, v, \theta, w) := \kappa \int_{\Gamma_{CARS}} w_i^2 \, \sigma \, do.$$

In chemical flow reactors **C**oherent **A**ntistokes **R**aman **S**pectroscopy (CARS) is used to produce signals which reflect the distributions of the mole fractions of certain reaction products $w_i$ along the measurement line $\Gamma_{CARS}$. The mole fraction of the species $w_i$ is determined by the balance equation

$$c_p \partial_t w_i - \nabla(\rho D_i \nabla w_i) + \rho v \cdot \nabla w_i = f_i(\theta, w),$$

together with equations for the other physical quantities, pressure $p$, velocity $v$, temperature $\theta$, and density $\rho$. The final goal is to determine certain reaction

velocities in the chemical process from these measurements. The accurate computation of the corresponding quantities is of crucial importance for this parameter
estimation process.

*Example* 1.2. *Mean surface pressure of a body in an inviscid flow* (Hartmann [71]):

$$J(\rho, v, e) := \int_S p \, do.$$

In the absence of viscosity, this quantity relates to the drag coefficient of the body.
The density $\rho$, the momentum $\rho v$ and the energy $e$ are determined by the Euler
equations, setting $p = (\gamma - 1)(e - \frac{1}{2}\rho v^2)$,

$$\partial_t \rho + \nabla \cdot (\rho v) = 0,$$
$$\partial_t (\rho v) + \nabla \cdot (\rho v \otimes v) + \nabla p = \rho g,$$
$$c_p \partial_t (\rho e) + c_p \nabla \cdot (\rho e v + p v) = h.$$

*Example* 1.3. *Boundary mean stress of an elasto-plastic body* (Rannacher and
Suttmeier [119]),

$$J(u, \sigma) := \int_{\Gamma_D} n^T \sigma n \, do,$$

over the 'clamped' part $\Gamma_D$ of the boundary. The stress tensor $\sigma$ and displacement
$u$ are determined by the Lamé-Navier equations with a pointwise stress constraint,

$$-\nabla \cdot \sigma = f, \quad \sigma = C\varepsilon, \quad \varepsilon = \tfrac{1}{2}(\nabla u + \nabla u^T), \quad |\sigma| \leq \sigma_0.$$

*Example* 1.4. *Local intensity measurement of a seismic signal* (Bangerth [14]):

$$J(u) := \int_{T-\delta}^{T+\delta} u(x_{\text{obs}}, t) \, dt.$$

The displacement $u$ is determined by the acoustic or elastic wave equation

$$\rho \partial_t^2 u - \nabla \cdot (a \nabla u) = 0.$$

In applications the elastic coefficient $a$ has to be determined from measurements
of this kind at varying points $x_{\text{obs}}$ resulting in an inverse problem.

*Example* 1.5. *Observed light emission of a proto-stellar dust cloud* (Kanschat [93]):

$$J(u) := \int_{n \cdot \theta_{obs} \geq 0} u(x, \theta_{\text{obs}}, \lambda_0) \, do.$$

The radiation intensity $u$ is determined by the *radiative transfer equation*

$$r_\theta \cdot \nabla u + (\kappa + \mu) u = B - \int_{S^d} R(\theta, \theta') u \, d\theta'.$$

Due to the distance of the light-emitting object, the observer (located on a satel-
lite) measures only the mean value of the intensity integrated over that part of
the surface of the object seen by the observer, and at some fixed wave length $\lambda_0$.

*Example* 1.6. *Critical eigenvalue in hydrodynamic stability analysis* (Heuveline and Rannacher [77]):

$$J(p, v, \lambda) := \lambda^{crit}.$$

The triple $\{p, v, \lambda\}$ is determined by the eigenvalue problem of the Navier-Stokes equations linearized about some stationary base flow $\hat{v}$,

$$-\nu \Delta v + \hat{v} \cdot \nabla v + v \cdot \nabla \hat{v} + \nabla p = \lambda v, \quad \nabla \cdot v = 0.$$

*Example* 1.7. *Cost functional in an optimization problem* (Becker et al. [28]):

$$J(u, q) := J_{\text{cost}}(u, q).$$

The state variable $v$ and control variable $q$ are determined by

$$J_{\text{cost}}(u, q) \rightarrow \min, \quad \mathcal{A}(u) + Bq = 0.$$

The state equation may, for example, consist of the 'incompressible' Navier-Stokes equations containing some control parameter $q$ (e.g., boundary control) and the cost functional may be the drag coefficient which is to be minimized.

## 1.4 General concepts of error estimation

In the following, we will introduce some of the main concepts of the DWR method for 'goal-oriented' a posteriori error estimation within the framework of linear algebra. We will use the same concepts later for differential equations as well and use this simple example only to introduce the most important aspects.

### Traditional error estimation

For regular matrices $A, A_h \in \mathbb{R}^{n \times n}$, and vectors $b, b_h \in \mathbb{R}^n$, consider the problems of finding $x, x_h \in \mathbb{R}^n$ from

$$Ax = b, \qquad A_h x_h = b_h, \tag{1.1}$$

where $h$ is a parameter indicating the quality of approximation, i.e., $A_h \rightarrow A$ and $b_h \rightarrow b$, as $h \rightarrow 0$. In this context, we introduce the notation of the *approximation error* $e := x - x_h$, the *truncation error* $\tau := A_h x - b_h$, and the *residual* $\rho := b - Ax_h$. Usually, *a priori* error analysis is based on the truncation error, and uses the identity

$$A_h e = A_h x - A_h x_h = A_h x - b_h = \tau,$$

to derive an a priori error bound involving a 'discrete' stability constant $c_{S,h}$:

$$\|e\| \leq c_{S,h} \|\tau\|, \qquad c_{S,h} := \|A_h^{-1}\|. \tag{1.2}$$

In contrast to that, the *a posteriori* error analysis uses the relation

$$Ae = Ax - Ax_h = b - Ax_h = \rho,$$

to derive an a posteriori error bound involving a 'continuous' stability constant:

$$\|e\| \le c_S \|\rho\|, \qquad c_S := \|A^{-1}\|. \tag{1.3}$$

Notice that the *a priori* error analysis is based on assumptions on the stability properties of the 'discrete' operator $A_h$, which may be difficult to establish for the particular approximation, while the *a posteriori* error analysis uses stability properties of the unperturbed 'continuous' operator $A$ which are often available from regularity theory. Further, the truncation error $\tau$ is not so easily computable in practical applications.

## Duality-based error estimation

In order to avoid the aforementioned drawbacks and to estimate the error also with respect to arbitrary moments of the solution, we employ a 'duality argument' well-known from the error analysis of Galerkin methods. For some given $j \subset \mathbb{R}^n$ assume that we want to estimate the value of the linear error functional

$$J(e) = J(u) - J(x_h) = (e, j).$$

For the determination of this error, consider the solution $z \in \mathbb{R}^n$ of the associated *dual* (or *adjoint*) *problem*

$$A^* z = j. \tag{1.4}$$

This leads us to an identity about the error

$$J(e) = (e, j) = (e, A^* z) = (Ae, z) = (\rho, z),$$

and finally to the 'weighted' a posteriori error estimate

$$|J(e)| \le \sum_{i=1}^{n} |\rho_i|\,|z_i|. \tag{1.5}$$

In this estimate the residuals $\rho_i$ are easily computable but the determination of the weights $z_i$ requires the solution of the auxiliary problem (1.4). The gain in using these weights is that they tell us about the influence of the 'local' residuals $\rho_i$ on the error in the target quantity $J(u)$.

We want to extend the concept of duality-based a posteriori error analysis to non-linear problems. Let differentiable mappings $A(\cdot)$, $A_h(\cdot) : \mathbb{R}^n \to \mathbb{R}^n$ and vectors $b$, $b_h \in \mathbb{R}^n$ be given and consider the problems

$$A(x) = b, \qquad A_h(x_h) = b_h. \tag{1.6}$$

Using the Jacobi matrix $A'(x)$ and the nonlinear residual $\rho := b - A(x_h)$, we have the following identities, for an arbitrary $y \in \mathbb{R}^n$ :

$$(A(x) - A(x_h), y) = \int_0^1 (A'(x_h + se)e, y) \, ds = (Be, y),$$

with the matrix

$$B := B(x, x_h) := \int_0^1 A'(x_h + se) \, ds.$$

For any given error functional $J(\cdot) = (\cdot, j)$, let $z \in \mathbb{R}^n$ be the solution of the corresponding dual problem

$$B^* z = j,$$

where regularity of $B^*$ is assumed. We obtain the error identity

$$J(e) = (e, j) = (e, B^* z) = (Be, z) = (A(x) - A(x_h), z) = (\rho, z),$$

which leads us to the weighted a posteriori error estimate

$$|J(e)| \leq \eta := \sum_{i=1}^n |\rho_i| \, |z_i|. \tag{1.7}$$

The use of the above error estimate requires the evaluation of the weights $|z_i|$. However, $z$ cannot be computed since it depends on the (unknown) error $e$ through the definition of $B$. To this end, we approximate the matrix $B$ by

$$B(x, x_h) \approx \tilde{B} := B(x_h, x_h) = \int_0^1 A'(x_h) \, ds = A'(x_h),$$

and solve the linearized dual problem, i.e. in practice an approximation of that,

$$A_h'(x_h)^* \tilde{z}_h = j_h.$$

This leads us to the approximate error estimate

$$|J(e)| \approx \tilde{\eta} := |(\rho, \tilde{z}_h)| \leq \sum_{i=1}^n \tilde{\omega}_i |\rho_i|, \qquad \tilde{\omega}_i := |\tilde{z}_{h,i}|. \tag{1.8}$$

The error by this approximation can be estimated. With $\tilde{z}$ the solution of

$$\tilde{B}^* \tilde{z} = A'(x_h)^* \tilde{z} = j,$$

we have by definition:

$$
\begin{aligned}
|J(e)| &= |(e, \tilde{B}^* \tilde{z})| = |(\tilde{B}e, \tilde{z})| \\
&\leq |((\tilde{B} - B)e, \tilde{z})| + |(Be, \tilde{z})| \\
&\leq |((\tilde{B} - B)e, \tilde{z})| + |(\rho, \tilde{z})| \\
&\leq |((\tilde{B} - B)e, \tilde{z})| + |(\rho, \tilde{z} - \tilde{z}_h)| + \tilde{\eta}.
\end{aligned}
$$

Using the estimate,

$$\|\tilde{B} - B\| = \left\| \int_0^1 \{A'(x_h) - A'(x_h + se)\} \, ds \right\| \leq \tfrac{1}{2} L' \|e\|,$$

with the Lipschitz constant $L'$ of $A'$, we obtain

$$|J(e)| \leq \tilde{\eta} + \tfrac{1}{2} L' \|e\|^2 \|\tilde{z}\| + \|\rho\| \, \|\tilde{z} - \tilde{z}_h\|. \tag{1.9}$$

Hence, assuming that $\|\tilde{z}\|$ can be controlled, the approximate error estimator $\tilde{\eta}$ is the dominant error term. However, notice that the derivation of (1.9) is based on typical 'linear algebra' arguments as it does not observe a possible dimension-dependence of norms and constants. In the PDE context, considered later on, we will have to argue more carefully in estimating the effect of linearization and approximation of the dual problem.

## Overview

The further contents of these Lecture Notes are as follows: In Chapter 2, we apply the principle of duality-based error estimation to the Galerkin approximation of ODEs. The following Chapter 3 is one of the core parts of these Lecture Notes. There, we develop the DWR method for the Poisson equation as the prototype of an elliptic problem. Then, Chapter 4 is devoted to the practical realization of error estimation and mesh adaptation. So far, the development of the DWR method is largely based on heuristic grounds, though strongly supported by computational experience. In Chapter 5, we discuss some of the central theoretical questions related to the justification of this approach but quickly get to the limits of theoretical analysis. Chapter 6 is the second core part of these Lecture Notes. It presents an abstract version of the DWR method for general nonlinear variational problems. This abstract approach is used in Chapter 7 for developing an a posteriori error analysis for the Galerkin approximation of eigenvalue problems. Another application is presented in Chapter 8, where the Galerkin approximation of optimization problems with PDE constraints is considered using the classical Lagrangian formalism. In Chapter 9, we realize the DWR method for the space-time discretization of nonstationary problems, with the heat and wave equations as model cases. Chapter 10 deals with the applications of the DWR method to linear and nonlinear problems from Structural Mechanics and Chapter 11 contains various results on the approximation of the incompressible Navier-Stokes equations as one of the basic models in fluid mechanics. In the last Chapter 12, some current developments and open problems are addressed.

# Chapter 2

# An ODE Model Case

In the following, we consider the realization of the ideas sketched in the Introduction for the initial value problem of an autonomous ODE system:

$$u'(t) = f(u(t)), \quad t \in I := [0, T], \quad u(0) = u_0. \tag{2.1}$$

We assume that the function $f(\cdot)$ is Lipschitz continuous and that the solution $u$ exists on the interval $[0, T]$. Very often, in applications, the goal in numerically approximating the solution is to know the end-time value

$$J(u) = u(T).$$

This goal may be reached by using the traditional *Finite Difference* (FD) method or the Galerkin *Finite Element* (FE) method with suitable error control and step-size selection strategies. The material of this chapter is mainly taken from Böttcher [37], and Böttcher and Rannacher [38].

## 2.1 Finite differences and finite elements

We start with a brief outline of the main philosophies underlying the Finite Difference and the Galerkin Finite Element schemes. Both types of methods are defined on time grids on the interval $I$ described by

$$0 = t_0 < \ldots < t_n < \ldots < t_N = T, \quad I_n = (t_{n-1}, t_n], \quad k_n = t_n - t_{n-1}.$$

### The finite difference method

We exemplarily consider the (implicit) *backward Euler scheme*: Find $U_n \sim u_n := u(t_n)$, such that

$$U_0 = u_0, \qquad U_n = U_{n-1} + k_n f(U_n) \quad n \geq 1. \tag{2.2}$$

Assuming the step size $k_n$ to be sufficiently small (or $f(\cdot)$ to be negative mono-tone), the discrete solution $U_n$ exists for all $n = 1, \ldots, N$. The corresponding truncation error

$$\tau_n(u) := k_n^{-1}(u_n - u_{n-1}) - f(u_n),$$

is bounded like

$$\|\tau_n(u)\| \leq \tfrac{1}{2} k_n \max_{I_n} \|u''\|.$$

Viewing the error $e_n = u_n - U_n$ as particular solution of the discrete scheme with right-hand side $\tau_n$, the discrete Gronwall lemma yields the usual a priori error estimate

$$\|e_N\| \leq K \sum_{n=1}^{N} k_n \|\tau_n(u)\|, \qquad K \sim \exp\Big(\int_0^T L_f(t)\,dt\Big), \tag{2.3}$$

where $L_f(t)$ is the Lipschitz constant of $f(\cdot)$ along the exact solution $u$. On this basis, we may use the following *explicit* formula for step-size control:

$$k_n = \frac{\text{TOL}}{KT\|\tau_n^0(U)\|}, \tag{2.4}$$

where the following approximation of the truncation error is used:

$$\tau_n(u) = k_n \tau_n^0(u) + \mathcal{O}(k_n^2) \approx k_n \tau_n^0(U).$$

*Remark* 2.1. The *a priori* error estimate (2.3) suffers from two short-comings:

- Usually the growth factor $K = K(T, L_f)$ is unknown as it depends on the exact solution.

- The step size formula requires an estimate for the 'leading' truncation error $\tau_n^0(u)$ along the exact solution. This, however, has to be generated (for example by local $h$-extrapolation) from the computed solution $U_n$.

## The Galerkin finite element method

We consider the approximation by the so-called *discontinuous Galerkin dG(0) (finite element) method*. Here, the space

$$S_h^{(0)}(I) := \{\varphi : I \to \mathbb{R}^d, \varphi_{|I_n} \in P_0(I_n)\},$$

consisting of piecewise constant functions, is used as 'trial' and 'test' space. The 'Galerkin approximation' $U \in S_h^{(0)}$ is determined by requiring that $U_0^- := u_0$ and

$$\sum_{n=1}^{N} \Big\{ \int_{I_n} (U' - f(U), \psi)\,dt + ([U]_{n-1}, \psi_{n-1}^+) \Big\} = 0 \quad \forall \psi \in S_h^{(0)}. \tag{2.5}$$

Here, we have used the standard notation

$$\varphi_n^- := \lim_{t \uparrow t_n} \varphi(t), \quad \varphi_n^+ := \lim_{t \downarrow t_n} \varphi(t), \quad [\varphi]_n := \varphi_n^+ - \varphi_n^-,$$

for left- and right-sided limits, and jumps of possibly discontinuous functions. Clearly, the continuous solution $u$ also satisfies the variational equation (2.5), and for the error $e := u - U$, there holds the nonlinear *Galerkin orthogonality* relation

$$\sum_{n=1}^{N} \left\{ \int_{I_n} (e' - f(u) + f(U), \psi) \, dt + ([e]_{n-1}, \psi_{n-1}^+) \right\} = 0 \quad \forall \psi \in S_h^{(0)}. \tag{2.6}$$

Since the test functions are allowed to be discontinuous, the globally formulated dG(0) method reduces to a time-stepping scheme which, in the present 'autonomous' case, is actually equivalent to the backward Euler scheme for the values $U_n := U_n^-$ :

$$U_0 = u_0, \qquad U_n - U_{n-1} = k_n f(U_n), \quad n \geq 1.$$

Now, the a posteriori error analysis for the end-time error $\|e_N\|$ via duality argument proceeds as follows. We consider the (backward in time) dual problem

$$-z' - B(t)^* z = 0, \quad T \geq t \geq 0, \qquad z(T) = \|e_N\|^{-1} e_N, \tag{2.7}$$

with the operator

$$B(t) := \int_0^1 f_x'(U + se) \, ds,$$

or in weak formulation,

$$\sum_{n=1}^{N} \left\{ \int_{I_n} (\varphi, -z' - B^* z) \, dt - (\varphi_n^-, [z]_n) \right\} = 0. \tag{2.8}$$

Taking $\varphi := e$, and using integration by parts and Galerkin orthogonality, we conclude that, with an arbitrary $Z \in S_h^{(0)}$,

$$\|e_N\| = \sum_{n=1}^{N} \int_{I_n} (e, -z' - B^* z) \, dt - \sum_{n=1}^{N-1} (e_n^-, [z]_n) + (e_N^-, z_N^-)$$

$$= \sum_{n=1}^{N} \int_{I_n} (e' - Be, z) \, dt + \sum_{n=2}^{N} ([e]_{n-1}, z_{n-1}^+) + (e_0^+, z_0^+)$$

$$= \sum_{n=1}^{N} \left\{ \int_{I_n} (e' - f(u) + f(U), z) \, dt + ([e]_{n-1}, z_{n-1}^+) \right\}$$

$$= \sum_{n=1}^{N} \left\{ \int_{I_n} (f(U), z - Z) \, dt - ([U]_{n-1}, (z - Z)_{n-1}^+) \right\}.$$

For the special choice $Z = \bar{z}$, the interval-wise mean value of $z$,

$$\bar{z}_{|I_n} := k_n^{-1} \int_{I_n} z \, dt,$$

we obtain the error representation

$$\|e_N\| = -\sum_{n=1}^{N} ([U]_{n-1}, (z - \bar{z})_{n-1}^+).$$

From this, we infer the following two types of a posteriori error estimates.

- Local 'weighted' a posteriori error estimate:

$$\|e_N\| \leq c_I \sum_{n=1}^{N} \left\{ k_n \|k_n^{-1}[U]_{n-1}\| \int_{I_n} \|z'\| \, ds \right\} =: c_I \sum_{n=1}^{N} k_n \rho_n \omega_n, \qquad (2.9)$$

  with the 'interpolation constant' $c_I = \frac{1}{2}$. The weights $\omega_n = \int_{I_n} \|z'\| \, ds$ represent the sensitivity of the error $\|e_N\|$ with respect to the local residuals $\rho_n = \|k_n^{-1}[U]_{n-1}\|$.

- Global a posteriori error estimate:

$$\|e_N\| \leq c_I \left( \max_{1 \leq n \leq N} \{k_n \rho_n\} \right) \int_I \|z'\| \, dt =: c_I c_S \max_{1 \leq n \leq N} \{k_n \rho_n\}. \qquad (2.10)$$

  The stability constant $c_S = \int_I \|z'\| \, dt$ represents the 'global' sensitivity of the error $\|e_N\|$ with respect to the maximal residual.

On the basis of the above estimates, we obtain the following 'implicit' a posteriori step-size control strategies ($k_n$ the old and $k_n'$ the new step size):

$$c_I N k_n \rho_n \omega_n \quad \text{or} \quad c_I c_S k_n \rho_n \left\{ \begin{array}{lcl} \geq \text{TOL} & \Rightarrow & k_n' = \frac{1}{2} k_n \\ \sim \text{TOL} & \Rightarrow & k_n' = k_n \\ \leq \frac{1}{4} \text{TOL} & \Rightarrow & k_n' = 2 k_n \end{array} \right\}.$$

Here, the weights $\omega_n$ and the stability constant $c_S$ can be approximated by the following formula, given a numerical approximation $Z$ of $z$:

$$\omega_n := \int_{I_n} \|z'\| \, dt \approx \|[Z]_n\|, \qquad c_S := \int_I \|z'\| \, dt \approx \sum_{n=1}^{N} \|[Z]_n\|.$$

*Remark* 2.2. The *a posteriori* error estimates (2.9) and (2.10) also have drawbacks which should be pointed out:

- The evaluation of the weights $\omega_n$ requires the solution of the dual problem over the whole time interval $[0, T]$.

- This 'global' step size control is 'implicit' since it involves the simultaneous adaptation of all time intervals $\{I_1, \ldots, I_N\}$ in each adaptation cycle.

*Remark* 2.3. Duality-based a posteriori error analysis for the Galerkin approximation of ODEs has also been considered in Estep and French [59], and Estep [58]. Here, the error estimates contain global stability constants which are derived by analytical arguments.

## 2.2 Efficiency comparison: FD versus FE method

In the following, we want to compare the efficiency of the step-size selection strategies described above for the FD and the FE method. We emphasize that in the present situation ('autonomous' ODE), both methods are only different ways of writing the very same scheme. Nevertheless, we will see that using either the FD or the FE approach gives quite different results. Consider the special scalar initial value problem

$$u'(t) = u(t)^2, \quad 0 \le t \le T < 1, \quad u(0) = 1,$$

with the singular solution (see Figure 2.1)

$$u(t) = \frac{1}{1-t}.$$

Let the goal of the computation again be the approximation of the end-time value $u(T)$. For the different mesh adaptation strategies designed for the backward Euler scheme and the dG(0) method, we want to estimate the work which is required for computing with accuracy TOL, in dependence on $T \to 1$.

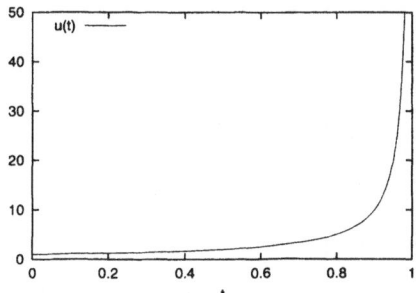 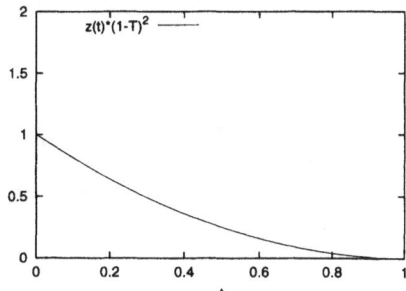

Figure 2.1: *Singular solution (left) and corresponding dual solution (right) for the evaluation of the end-time error.*

## The dG(0) method

Let the step-size control be based on the *global* a posteriori error estimate (2.10),

$$|e_N| \leq c_I c_S \max_{1 \leq n \leq N} \{k_n \rho_n\}, \qquad \rho_n := k_n^{-1} |[U]_{n-1}|,$$

with the interpolation constant $c_I = \frac{1}{2}$ and the stability constant $c_S$ from the dual a priori bound

$$\int_0^T |z'| \, dt =: c_S.$$

The criterion for the choice of the local step size is

$$c_I c_S k_n \rho_n = \text{TOL} \quad \Leftrightarrow \quad k_n = \frac{\text{TOL}}{c_I c_S \rho_n} \quad \Rightarrow \quad |e_N| \leq \text{TOL}.$$

Notice that, by construction, $\sum_{n=1}^{N} k_n = T$. For the explicit estimation of the work count, the parameters in this formula are approximated as follows:
i) Suppose that the step size is already small enough such that

$$\rho_n = k_n^{-1} |[U]_n| \approx k_n^{-1} |u_n - u_{n-1}| \approx \sup_{I_n} |u'| \approx (1-t_n)^{-2}.$$

ii) The stability constant $c_S$ is essentially determined by the dual problem (after linearization about the continuous solution $u$)

$$z'(t) = -\frac{2}{1-t} z(t), \quad 1 > T \geq t \geq 0, \quad z(T) = 1,$$

with the solution

$$z(t) = \exp \left( 2 \int_t^T \frac{ds}{1-s} \right) z(T) = \left( \frac{1-t}{1-T} \right)^2.$$

Hence,

$$c_S = \int_0^T |z'| \, dt \leq \frac{1}{(1-T)^2}.$$

This leads us to the step-size distribution

$$k_n \approx (1-t_n)^2 (1-T)^2 \, \text{TOL}.$$

The corresponding work measure $N$ is determined as

$$N = \sum_{n=1}^{N} k_n k_n^{-1} \approx \frac{1}{(1-T)^2 \text{TOL}} \sum_{n=1}^{N} k_n (1-t_n)^{-2} \approx \frac{1}{(1-T)^2 \text{TOL}} \int_0^T \frac{dt}{(1-T)^2},$$

and consequently

$$N \approx \frac{1}{(1-T)^3} \frac{1}{\text{TOL}}.$$ (2.11)

For later purposes, we note that the local weights are determined by

$$\omega_n = \int_{I_n} |z'(t)| \, dt \approx k_n \frac{1-t_n}{(1-T)^2}.$$

## The backward Euler scheme

Let the step-size control be based on the *global* a priori error estimate

$$|e_N| \leq K \sum_{n=1}^{N} k_n |\tau_n(u)|, \qquad |\tau_n(u)| \leq \tfrac{1}{2} k_n \sup_{I_n} |u''|.$$

The corresponding criterion for the control of the local step sizes is

$$k_n = \frac{\text{TOL}}{KT \sup_{I_n} |u''|} \qquad \Rightarrow \qquad |e_N| \leq \text{TOL}.$$

The parameters in this formula are approximated as follows:
i) The Lipschitz constant of $f(\cdot)$ along the solution $u$ is

$$L_f(t) = \frac{2}{1-t},$$

such that the growth factor becomes

$$K(T, L_f) \approx \exp\left(\int_0^T L_f(t) \, dt\right) \approx \exp\left(\int_0^T \frac{2}{1-t} \, dt\right) \approx (1-T)^{-2}.$$

ii) The leading truncation error $|\tau_n^0|$ behaves like

$$|\tau_n^0| \leq \tfrac{1}{2} \sup_{I_n} |u''| \approx (1-t_n)^{-3}.$$

This results in the step-size distribution

$$k_n \approx \frac{\text{TOL}}{TK(T) \sup_{I_n} |u''|} \approx (1-T)^2 (1-t_n)^3 \, \text{TOL}.$$

The corresponding work count is

$$N = \sum_{n=1}^{N} k_n k_n^{-1} \approx \frac{1}{(1-T)^2} \frac{1}{\text{TOL}} \sum_{n=1}^{N} k_n (1-t_n)^{-3},$$

and consequently (compare this to (2.11)),

$$N \approx \frac{1}{(1-T)^4} \frac{1}{\text{TOL}}.$$ (2.12)

*Remark* 2.4. We note that for both FD-based and FE-based schemes the efficiency of the computation can be further improved by using a more localized adaptation strategy. This will be considered in the exercises below.

*Remark* 2.5. The critical drawback of the heuristic step-size control strategy used for the backward Euler scheme is the possible crude under-estimation of the growth factor $K(T, L_f)$. On the other hand, estimating $K(T, L_f)$ by an *a priori* analysis is oriented at the *worst case* scenario and usually leads to over-estimation rendering the resulting error bound useless. In general, this prevents the step-size control strategy based on the *a priori* error estimate (2.3) from providing a real *control* of the error. In order to overcome this limitation the following two different approaches may be used:

i) The first one is based on the relation

$$e_n = e_{n-1} + k_n f'(U_n)e_n + k_n \tau_n(u) + k_n O(e_n^2), \qquad (2.13)$$

for the error $e_n = u_n - U_n$, with initial value $e_0 = 0$. Using a guess for the truncation error $\tau_n(U_n) \approx \tau_n(u)$, obtained for example by local extrapolation, the solution $E_n$ of the *linearized* error equation

$$E_n = E_{n-1} + k_n f'(U_n)E_n + k_n \tau_n(U_n), \quad 0 \le n \le N, \qquad (2.14)$$

may be used to get a guess for the true end-time error, $E_N \approx e_N$. However, this does not provide criteria for a time-step adaptation to reduce the error below the prescribed tolerance, $\|e_N\| \le TOL$.

ii) An alternative approach employs a duality argument similar to that one used for the dG(0) method, but now on the *discrete* level. Let $Z_n$ be the solution of the linearized backward-in-time scheme

$$Z_{n-1} = Z_n + k_n f'(U_n)Z_{n-1}, \quad 0 \le t_n < t_N. \qquad (2.15)$$

with starting value $Z_N$. Then, using the error relation (2.13), we obtain

$$
\begin{aligned}
(e_n, Z_n) &= (e_n, Z_n - Z_{n-1}) + (e_n - e_{n-1}, Z_{n-1}) + (e_{n-1}, Z_{n-1}) \\
&= -k_n(e_n, f'(U_n)Z_{n-1}) + k_n(f'(U_n)e_n, Z_{n-1}) \\
&\quad + k_n(\tau_n(u) + \mathcal{O}(e_n^2), Z_{n-1}) + (e_{n-1}, Z_{n-1}),
\end{aligned}
$$

and summing over $1 \le n \le N$,

$$(e_N, Z_N) = (e_0, Z_0) + \sum_{n=1}^{N} k_n(\tau_n(u) + \mathcal{O}(e_n^2), Z_{n-1}). \qquad (2.16)$$

For $Z_N := e_N \|e_N\|^{-1}$ and $e_0 = 0$, we arrive at

$$\|e_N\| \le c_{S,k} \sum_{n=1}^{N} k_n \{ \|\tau_n(u)\| + \mathcal{O}(\|e_n\|^2) \}, \qquad (2.17)$$

with the *discrete* stability constant $c_{S,k} := \max_{0 \le n \le N-1} \|Z_n\|$. Neglecting the quadratic error term, we obtain the error estimate

$$\|e_N\| \approx c_{S,k} \sum_{n=1}^{N} k_n \|\tau_n(U_n)\|, \qquad (2.18)$$

with an approximation $\tau_n(U_n) \approx \tau_n(u)$ for the truncation error. Here, the generally too pessimistic *a priori* bound $K(t_N, L_f)$ is replaced by the *a posteriori* stability constant $c_{S,k}$. In practice, $c_{S,k}$ will be determined from the computed dual solution $Z_n$ using again a suitable guess for the starting value $Z_N = e_N \|e_N\|^{-1}$. The step-size selection strategy based on the estimate (2.18), is generically *implicit* like that proposed for the dG(0) method, but is capable of producing useful error bounds. Surprisingly, it has not found much attention in the FD community.

## 2.3  Exercises

*Exercise* 2.1. Consider the autonomous initial value problem

$$u'(t) = u(t)^2, \quad 0 \le t \le T < 1, \quad u(0) = 1,$$

with the solution $u(t) = (1-t)^{-1}$. Compute $u(T)$ by the backward Euler scheme

$$U_n = U_{n-1} + k_n U_n^2, \quad n \ge 1, \quad U_0 = 1.$$

Let the step sizes $k_n$ be chosen on the basis of the a priori error bound (2.3),

$$|e_N| \le K(T) \sum_{n=1}^{N} k_n \tau_n(u), \qquad \tau_n(u) = k_n \tau_n^0(u) + \mathcal{O}(k_n^2),$$

according to the *implicit* control

$$k_n \approx \left( \frac{\text{TOL}}{N \, K(T) |\tau_n^0(u)|} \right)^{1/2}.$$

What is the asymptotic work count $N = N(\text{TOL}, T)$ as $T \to 1$?

*Exercise* 2.2. Consider the model problem of Exercise 2.1. Let in the dG(0) method the step-size selection be based on the *weighted* error estimator (2.9),

$$|e_N| \le c_I \sum_{n=1}^{N} k_n \rho_n \omega_n,$$

according to the *implicit* control

$$k_n \approx \frac{\text{TOL}}{N \rho_n \omega_n}.$$

What is the work count $N = N(\text{TOL}, T)$ as $T \to 1$?

*Exercise* 2.3. The dG(0) method is only of first order. By the same principle, implicit dG(r) methods of any order $r \geq 1$ can be designed. Alternatives are the so-called *cG(r) (continuous Galerkin) methods* which are Petrov-Galerkin methods. The simplest representative is the second-order cG(1) method which uses (continuous) piecewise linear trial and (discontinuous) piecewise constant test functions:

$$U(0) = u_0, \quad \sum_{n=1}^{N} \int_{I_n} (U' - f(U), \psi) \, dt = 0 \qquad \forall \psi \in S_h^{(0)}(I).$$

This method is closely related to the implicit midpoint rule

$$U_0 = u_0, \quad U_n = U_{n-1} + k_n f(\tfrac{1}{2}\{U_{n-1} + U_n\}), \quad n \geq 1.$$

Develop a residual-based a posteriori error estimate for the end-time error $\|e_N\|$ of the cG(1) method.

*Exercise* 2.4 *(Practical exercise).* Verify the theoretical predictions in Exercises 2.1 and 2.2 by a computational experiment. Use the backward Euler scheme and the dG(0) method for approximating the solution value $u(T)$ in the model problem

$$u'(t) = u(t)^2, \quad 0 \leq t \leq T < 1, \quad u(0) = 1,$$

with the solution $u(t) = (1-t)^{-1}$.

a) Use all the step-size selection strategies developed in the text and in Exercises 2.1 and 2.2 with the formulas for $K(T)$, $|\tau_n^0(u)|$, $\rho_n$ and $\omega_n$ as given in the text. Determine experimentally the number of resulting time steps $N$ for a decaying sequence of tolerances $\mathrm{TOL}_i = 2^{-i}$, $i = 1, 2, 3, \ldots$. Monitor the true error $e(T) = u(T) - U_N$ and compare it with the given tolerance. Interpret the observed results.

b) Approximate the leading truncation errors $|\tau_n^0|$ by second-order difference quotients of the computed solution $U_n$, and the residuals $\rho_n = k_n^{-1}|[U]_n|$, as defined in the text. Repeat the above test using these approximations and report the differences to the results observed in (a).

c) Do the same test as in (a) for a sequence of *uniform* step size distributions with $k \equiv T/N$. You will observe that the performance of equidistant time meshes is almost as good as that of the best adaptive procedure based on the 'weighted' a posteriori error estimate for the dG(0) method. This surprising phenomenon can be explained by showing that in this case

$$N \approx (1-T)^{-2} |\log(1-T)| \, \mathrm{TOL}^{-1}.$$

Try to detect this logarithmic behavior in the numerical experiment. So what is the benefit of sophisticated local time-step adaptation?

d) Perform the same experiment with the cG(1) method described in Exercise 2.3. First, use uniform step sizes and, then, try to use your a posteriori error indicator derived in Exercise 2.3. This should demonstrate the superiority of the second-order cG(1) over the only first-order dG(0) method.

# Chapter 3

# A PDE Model Case

In this chapter, we will develop the basics of the DWR method for linear elliptic partial differential equations as originally described in Becker and Rannacher [30]. As a model configuration, we consider the Poisson equation on a polygonal or polyhedral domain $\Omega \subset \mathbb{R}^d$, with Dirichlet boundary conditions:

$$-\Delta u = f \quad \text{in } \Omega, \quad u_{|\partial\Omega} = 0. \tag{3.1}$$

The discretization will be by a Galerkin finite element method which is based on the variational formulation of (3.1): Find $u \in V := H_0^1(\Omega)$, such that

$$a(u, \psi) := (\nabla u, \nabla \psi) = (f, \psi) \quad \forall \psi \in V. \tag{3.2}$$

In this chapter, we will exemplarily consider the a posteriori control of the resulting approximation $u_h$ with respect to the following goals:

- Computation of an overview of the solution's structure (global norm error):

$$\|\nabla(u - u_h)\| \leq TOL, \qquad \|u - u_h\| \leq TOL.$$

- Computation of 'displacement' or 'stress' components at some point $a \in \bar{\Omega}$ (point-value error):

$$J(u) := u(a), \qquad J(u) := \partial_i u(a).$$

- Computation of 'mean normal flux':

$$J(u) := \int_{\partial\Omega} \partial_n u \, ds.$$

However, we will always have in mind the accurate computation with respect to *arbitrary* functional values $J(u)$ of the solution.

Here and below, we use the following notation: For a domain $\Omega \subset \mathbb{R}^d$, $L^2(\Omega)$ is the Lebesgue space of square-integrable functions on $\Omega$, which is a Hilbert space with the scalar product and norm

$$(v, w)_\Omega = \int_\Omega v\, w\, dx, \qquad \|v\|_\Omega = \left( \int_\Omega |v|^2\, dx \right)^{1/2}.$$

Analogously, $L^2(\partial\Omega)$ is the space of square-integrable functions defined on the boundary $\partial\Omega$ equipped with the scalar product and norm

$$(v, w)_{\partial\Omega} = \int_{\partial\Omega} v\, w\, ds, \qquad \|v\|_{\partial\Omega} = \left( \int_{\partial\Omega} |v|^2\, ds \right)^{1/2}.$$

The Sobolev spaces $H^1(\Omega)$ and $H^2(\Omega)$ consist of those functions $v \in L^2(\Omega)$ which possess first- and second-order (distributional) derivatives $\nabla v \in L^2(\Omega)^d$ and $\nabla^2 v \in L^2(\Omega)^{d \times d}$, respectively. For functions in these spaces, we use the semi-norms

$$|v|_{1;\Omega} := \|\nabla v\|_\Omega, \qquad |v|_{2;\Omega} := \|\nabla^2 v\|_\Omega.$$

This notation can be extended to Sobolev spaces $H^p(\Omega)$ of arbitrary order $p \geq 1$. The space $H^1(\Omega)$ can be embedded in the space $L^2(\partial\Omega)$, such that for each $v \in H^1(\Omega)$ there exists a trace $v_{|\partial\Omega} \in L^2(\partial\Omega)$. Further, the functions in the subspace $H^1_0(\Omega) \subset H^1(\Omega)$ are characterized by the property $v_{|\partial\Omega} = 0$. By the Poincaré inequality,

$$\|v\|_\Omega \leq c \|\nabla v\|_\Omega, \quad v \in H^1_0(\Omega), \tag{3.3}$$

the $H^1$-semi-norm $\|\nabla v\|_\Omega$ is a norm on the subspace $H^1_0(\Omega)$. If the set $\Omega$ is identical with the domain on which the differential equation is posed, we usually omit the subscript $\Omega$ in the notation of norms and scalar products, for instance $\|v\| = \|v\|_\Omega$. All the above notation will be synonymously used also for vector- or matrix-valued functions $v : \Omega \to \mathbb{R}^d$ or $\mathbb{R}^{d \times d}$.

## 3.1   Finite element approximation

The discretization of the model problem (3.1) seeks an approximations $u_h \in V_h$, the so-called *Ritz projection* of $u$, in a certain finite dimensional subspace $V_h \subset V$,

$$a(u_h, \psi_h) = (f, \psi_h) \quad \forall \psi_h \in V_h. \tag{3.4}$$

The main feature of the Galerkin method for linear problems is the so-called *Galerkin orthogonality* for the error $e := u - u_h$,

$$a(e, \psi_h) = 0, \quad \psi_h \in V_h. \tag{3.5}$$

The subspaces (finite element spaces) considered have the form

$$V_h = \{v \in V : v_{|K} \in P(K), \ K \in \mathbb{T}_h\},$$

defined on decompositions $\mathbb{T}_h$ of $\Omega$ into *cells* $K$ (triangles or quadrilaterals in $\mathbb{R}^2$, and tetrahedra or hexahedra in $\mathbb{R}^3$) of width $h_K = \mathrm{diam}(K)$; we write $h = \max_{K \in \mathbb{T}_h} h_K$ for the *global* mesh width. Here, $P(K)$ denotes a suitable space of polynomial-like functions defined on the cell $K \in \mathbb{T}_h$. In the numerical results discussed below, we have mostly used 'bilinear' or 'trilinear' finite elements on quadrilateral or hexahedral meshes, respectively, in which case $P(K) = \tilde{Q}_1(K)$ consists of shape functions obtained via a bilinear transformation from the space of 'bilinears' $Q_1(\hat{K}) = \mathrm{span}\{1, x_1, x_2, x_1 x_2\}$ or 'trilinears' $Q_1(\hat{K}) = \mathrm{span}\{1, x_1, x_2, x_3, x_1 x_2, x_2 x_3, x_3 x_1, x_1 x_2 x_3\}$ on the reference cell $\hat{K} = [0,1]^d$. Local mesh refinement or coarsening is realized by using *hanging nodes* in such a way that global conformity is preserved, that is $V_h \subset V$ (see also Section 4.2). For technical details of finite element spaces, the reader may consult the standard literature, for instance Ciarlet [46], Johnson [83] or Brenner and Scott [43], and especially Carey and Oden [44] for the treatment of hanging nodes.

We consider the control of the error with respect to some 'output functional' $J(\cdot)$, i.e., we want to have estimates for the difference $J(e) = J(u) - J(u_h)$. For simplicity, we assume here $J(\cdot)$ to be linear. Following the general concept of the DWR method, let $z \in V$ be the solution of the associated *dual problem*

$$a(\varphi, z) = J(\varphi) \quad \forall \varphi \in V, \tag{3.6}$$

and $z_h \in V_h$ its finite element approximations defined by

$$a(\varphi_h, z_h) = J(\varphi_h) \quad \forall \varphi_h \in V_h. \tag{3.7}$$

Using this construction together with Galerkin orthogonality, we obtain

$$\begin{aligned}
J(e) &= a(e, z) = a(e, z - \psi_h) \\
&= (f, z - \psi_h) - a(u_h, z - \psi_h) =: \rho(u_h)(z - \psi_h), \quad \psi_h \in V_h.
\end{aligned}$$

The so-called *residual* $\rho(u_h)(\cdot)$ of the Galerkin approximation $u_h$ may be viewed as a functional on the solution space $V$. Cell-wise integration by parts implies

$$\begin{aligned}
\rho(u_h)(z - \psi_h) &= \sum_{K \in \mathbb{T}_h} \left\{ (f + \Delta u_h, z - \psi_h)_K - (\partial_n u_h, z - \psi_h)_{\partial K} \right\} \\
&= \sum_{K \in \mathbb{T}_h} \left\{ (f + \Delta u_h, z - \psi_h)_K + \tfrac{1}{2}([\partial_n u_h], z - \psi_h)_{\partial K \setminus \partial \Omega} \right\},
\end{aligned}$$

where $[\partial_n u_h]$ denotes the jump of $\partial_n u_h$ across the inter-element edges (in 2-D) or faces (in 3-D), i.e., for two neighboring cells $K, K' \in \mathbb{T}_h$ with common edge $\Gamma$ and normal unit vector $n$ pointing from $K$ to $K'$, we set

$$[\partial_n u_h] = [\nabla u_h \cdot n] := (\nabla u_{h|K' \cap \Gamma} - \nabla u_{h|K \cap \Gamma}) \cdot n.$$

Since $n = -n'$, this actually defines the jump of $\partial_n u_h$ across the edge $\Gamma$. For later use, we define the cell and edge residuals $R_h$ and $r_h$, respectively, by

$$R_{h|K} := f + \Delta u_h,$$

$$r_{h|\Gamma} := \begin{cases} \frac{1}{2}[\partial_n u_h], & \text{if } \Gamma \subset \partial K \backslash \partial \Omega, \\ 0, & \text{if } \Gamma \subset \partial \Omega. \end{cases}$$

We collect the previous results in the following Proposition.

**Proposition 3.1.** *For the finite element approximation (3.4) of the Poisson problem, we have the a posteriori error representation*

$$J(e) = \sum_{K \in \mathbb{T}_h} \left\{ (R_h, z - \psi_h)_K + (r_h, z - \psi_h)_{\partial K} \right\}, \tag{3.8}$$

*with an arbitrary $\psi_h \in V_h$, and as a consequence the a posteriori error estimate*

$$|J(e)| \leq \eta_\omega := \sum_{K \in \mathbb{T}_h} \rho_K \, \omega_K, \tag{3.9}$$

*where the cell residuals ('smoothness indicators') $\rho_K$ and weights ('influence factors') $\omega_K$ are given by*

$$\rho_K := \left( \|R_h\|_K^2 + h_K^{-1} \|r_h\|_{\partial K}^2 \right)^{1/2},$$

$$\omega_K := \left( \|z - \psi_h\|_K^2 + h_K \|z - \psi_h\|_{\partial K}^2 \right)^{1/2},$$

*for an arbitrary $\psi_h \in V_h$.*

*Remark* 3.2. The *dual solution* $z$ has the features of a 'generalized' Green function $G(K, K')$, as it describes the dependence of the target error quantity $J(e)$, which may be concentrated at some cell $K$, on local properties of the data, i.e. in this case the residuals $\rho_{K'}$ on cells $K'$; see Figure 3.1.

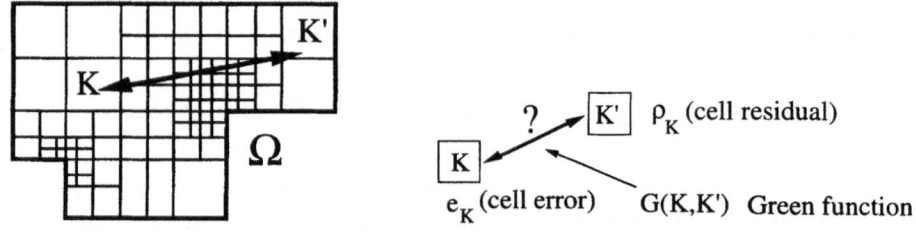

Figure 3.1: *Finite element mesh and scheme of error propagation.*

In order to evaluate the a posteriori error representation (3.8) or the resulting a posteriori error estimate (3.9), we need information about the 'continuous' dual solution $z$. Since in practice, $z$ is not explicitly known, such information has to be obtained either through *a priori* analysis in form of bounds for $z$ in certain Sobolev norms or through computation by solving the dual problem numerically. In the following, we will present some examples in which $z$ can be bounded *a priori* or can even be explicitly determined. In later sections, dealing with real-life problems, we will always have to rely on the computational approximation of $z$.

## 3.2 Global a posteriori error estimates

In the following, we will demonstrate how the previous approach can be used for deriving the known a posteriori error estimates with respect to 'global' norms.

*Example* 3.1. (*Energy-norm error*) For deriving the usual error estimate with respect to the natural *energy norm* associated with problem (3.2), we choose the error functional

$$J(\varphi) = (\nabla\varphi, \nabla e)\|\nabla e\|^{-1},$$

considering the error $e$ as a fixed quantity. The corresponding dual solution $z \in V$ is determined by

$$a(\varphi, z) = (\nabla\varphi, \nabla e)\|\nabla e\|^{-1} \quad \forall \varphi \in V,$$

and admits the trivial a priori bound $\|\nabla z\| \leq 1 =: c_S$ (stability constant). Clearly, in this particular case the dual solution is just given by $z = e\|\nabla e\|^{-1}$. From Proposition 3.1 we infer the estimate

$$J(e) = \|\nabla e\| \leq \sum_{K \in \mathbb{T}_h} \rho_K\, \omega_K \leq \Big( \sum_{K \in \mathbb{T}_h} h_K^2 \rho_K^2 \Big)^{1/2} \Big( \sum_{K \in \mathbb{T}_h} h_K^{-2} \omega_K^2 \Big)^{1/2}.$$

We recall the interpolation error estimate (see, e.g., Scott and Zhang [125])

$$\inf_{\psi_h \in V_h} \Big( \sum_{K \in \mathbb{T}_h} \big\{ h_K^{-2}\|z - \psi_h\|_K^2 + h_K^{-1}\|z - \psi_h\|_{\partial K}^2 \big\} \Big)^{1/2} \leq \tilde{c}_I \|\nabla z\|, \qquad (3.10)$$

to estimate further,

$$\|\nabla e\| \leq \tilde{c}_I \Big( \sum_{K \in \mathbb{T}_h} h_K^2 \rho_K^2 \Big)^{1/2} \|\nabla z\| \leq \tilde{c}_I \Big( \sum_{K \in \mathbb{T}_h} h_K^2 \rho_K^2 \Big)^{1/2}.$$

This results in the classical *energy-norm error estimate*

$$\|\nabla e\| \leq \eta_E := \tilde{c}_I \Big( \sum_{K \in \mathbb{T}_h} h_K^2 \rho_K^2 \Big)^{1/2}, \qquad (3.11)$$

with $\rho_K$ as defined in Proposition 3.1.

*Remark* 3.3. By Galerkin orthogonality, there holds

$$
\begin{aligned}
\|\nabla e\|^2 &= \|\nabla u\|^2 - 2(\nabla u, \nabla u_h) + \|\nabla u_h\|^2 \\
&= \|\nabla u\|^2 - 2(\nabla u_h, \nabla u_h) + \|\nabla u_h\|^2 = \|\nabla u\|^2 - \|\nabla u_h\|^2,
\end{aligned}
$$

such that *energy-norm error control* turns out to be equivalent to *energy error control*. However, this requires the energy form to be a scalar product.

*Example* 3.2. (*$L^2$-norm error*) To derive an estimate with respect to the $L^2$ norm, we choose the error functional

$$
J(\varphi) = (\varphi, e)\|e\|^{-1}.
$$

Suppose that the (polygonal or polyhedral) domain $\Omega$ is convex. Then, the corresponding dual solution $z \in V \cap H^2(\Omega)$ admits the a priori bound $\|\nabla^2 z\| \le 1 =: c_S$ (stability constant). From the result of Proposition 3.1, we infer the estimate

$$
\|e\| \le \sum_{K \in \mathbb{T}_h} \rho_K \,\omega_K \le \Big( \sum_{K \in \mathbb{T}_h} h_K^4 \rho_K^2 \Big)^{1/2} \Big( \sum_{K \in \mathbb{T}_h} h_K^{-4} \omega_K^2 \Big)^{1/2}.
$$

Using the interpolation error estimate (see, e.g., Brenner and Scott [43])

$$
\inf_{\psi_h \in V_h} \Big( \sum_{K \in \mathbb{T}_h} \{ h_K^{-4}\|z - \psi_h\|_K^2 + h_K^{-3}\|z - \psi_h\|_{\partial K}^2 \} \Big)^{1/2} \le c_I \|\nabla^2 z\|, \qquad (3.12)
$$

we obtain

$$
\|e\| \le c_I \Big( \sum_{K \in \mathbb{T}_h} h_K^4 \rho_K^2 \Big)^{1/2} \|\nabla^2 z\| \le c_I c_S \Big( \sum_{K \in \mathbb{T}_h} h_K^4 \rho_K^2 \Big)^{1/2}.
$$

This results in the well-known *$L^2$-norm error estimate*

$$
\|e\| \le \eta_{L^2} := c_I c_S \Big( \sum_{K \in \mathbb{T}_h} h_K^4 \rho_K^2 \Big)^{1/2}. \qquad (3.13)
$$

with $\rho_K$ again as defined in Proposition 3.1. In comparison to the energy-norm error estimate (3.11) the $L^2$-norm estimate involves the weighting $h_K^4$ which reflects its higher order of convergence.

## 3.3   A posteriori error estimates for output functionals

Next, we turn to estimating the error with respect to local error functionals. Let *TOL* denote the accuracy we want to achieve.

*Example* 3.3. (*Point-value error*) To estimate the error at some point $a \in \Omega$, we use the regularized functional

$$J(u) := |B_\varepsilon|^{-1} \int_{B_\varepsilon} u \, dx = u(a) + \mathcal{O}(\varepsilon^2),$$

where $B_\varepsilon$ is the $\varepsilon$-ball around the point $a$ and $\varepsilon := TOL$. The corresponding dual solution $z$ behaves like a regularized Green function, i.e. in 2-D:

$$z(x) = g_\varepsilon^a(x) \approx \log(r(x)), \quad r(x) := \sqrt{|x-a|^2 + \varepsilon^2}.$$

By choosing $\psi_h$ in the estimate (3.9) suitably, the weights have here the form

$$\omega_K \approx h_K^2 \|\nabla^2 z\|_K \approx h_K^2 |K|^{1/2} r_K^{-2}, \qquad r_K := \max_{x \in K} r(x),$$

such that

$$|e(a)| \approx \eta_\omega := c_I \sum_{K \in \mathbb{T}_h} \frac{h_K^3}{r_K^2} \rho_K. \tag{3.14}$$

*Example* 3.4. (*Point-value derivative error*) To estimate the error in the derivative in direction $x_i$ at some interior point $a \in \Omega$, we use the regularized output functional

$$J(u) := |B_\varepsilon|^{-1} \int_{B_\varepsilon} \partial_i u \, dx = \partial_i u(a) + \mathcal{O}(\varepsilon^2),$$

where again $\varepsilon := TOL$. In this case the dual solution behaves like a regularized derivative Green function, i.e. in 2-D,

$$z(x) = \partial_i g_\varepsilon^a(x) \approx \frac{(x-a)_i}{r(x)^2}, \quad r(x) := \sqrt{|x-a|^2 + \varepsilon^2},$$

and the corresponding weights like

$$\omega_K \approx h_K^2 \|\nabla^2 z\|_K \approx h_K^2 |K|^{1/2} r_K^{-3}.$$

This results in the a posteriori error estimate

$$|\partial_i e(a)| \approx \eta_\omega := c_I \sum_{K \in \mathbb{T}_h} \frac{h_K^3}{r_K^3} \rho_K. \tag{3.15}$$

Compared with (3.14), this localizes the region of influence towards the point $a$ even more.

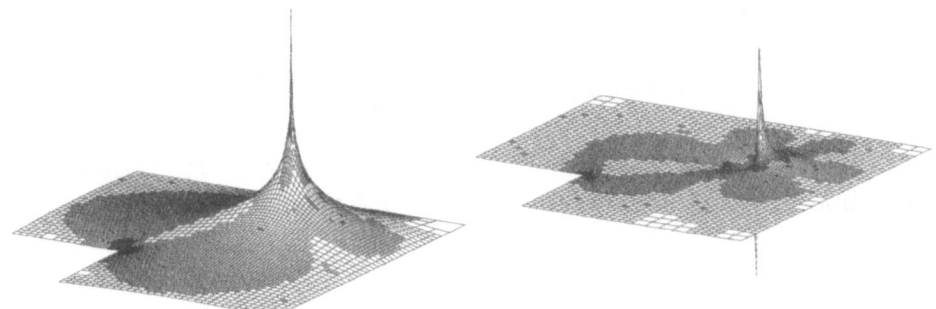

Figure 3.2: *Examples of computed dual solutions: regularized Green function and derivative Green function (scaled differently), for evaluating $u(a)$ and $\partial_1 u(a)$.*

*Example* 3.5. (*Mean normal-flux error*) Another type of error functionals involves integrals over lower-dimensional manifolds. As an example, we consider the computation of the mean normal flux across the boundary,

$$J(u) = \int_{\partial\Omega} \partial_n u \, ds,$$

where for simplicity $\Omega$ is assumed to be the unit circle. The question is: *What is an 'efficient' mesh-size distribution for computing* $J(u)$? Notice that in this simple context the computational goal is trivial since it can be reduced to evaluating data, $J(u) = \int_\Omega \Delta u \, dx = -\int_\Omega f \, dx$. However, in more complex situations error functionals of this type cannot be so easily computed and are of high interest (c.f. the drag coefficient or the average Nusselt number mentioned in the Introduction). Here, the corresponding dual problem

$$a(\varphi, z) = (1, \partial_n \varphi)_{\partial\Omega} \quad \forall \varphi \in V \cap C^1(\bar{\Omega})$$

has a measure solution of the type $z \equiv -1$ in $\Omega$, $z = 0$ on $\partial\Omega$. Hence, to avoid dealing with measures, we use the regularized output functional

$$J_\varepsilon(\varphi) = \varepsilon^{-1} \int_{S_\varepsilon} \partial_n \varphi \, dx = \int_{\partial\Omega} \partial_n \varphi \, ds + \mathcal{O}(\varepsilon),$$

where $S_\varepsilon = \{x \in \Omega : \operatorname{dist}\{x, \partial\Omega\} < \varepsilon\}$ and $\varepsilon := TOL$. The corresponding dual solution is explicitly given by

$$z_\varepsilon = \begin{cases} -1 & \text{in } \Omega \backslash S_\varepsilon, \\ -\varepsilon^{-1}\operatorname{dist}\{x, \partial\Omega\} & \text{in } S_\varepsilon. \end{cases}$$

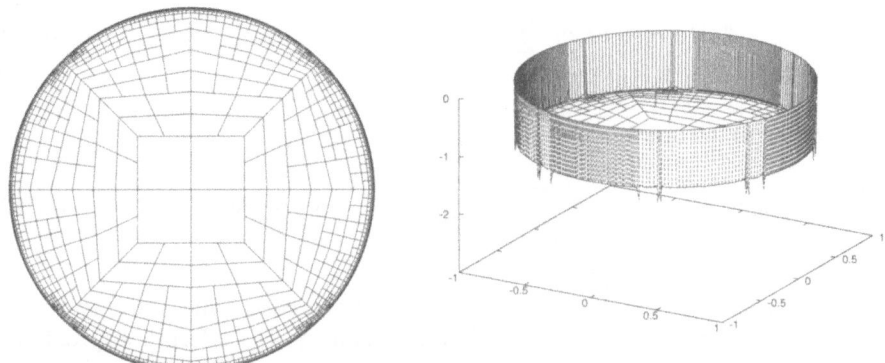

Figure 3.3: *Refined mesh and computed dual solution for the mean normal flux.*

On cells $K \subset \Omega \backslash S_\varepsilon$ there holds $z - I_h z \equiv 0$, which leads us to the error estimate

$$J_\varepsilon(e) \leq \eta_\omega := \sum_{K \in \mathbb{T}_h, K \cap S_\varepsilon \neq \emptyset} \rho_K \omega_K.$$

The conclusion is: *There is no contribution to the error from cells in the interior of* $\Omega$. Hence, whatever right-hand side $f$, the optimal strategy is to refine the elements adjacent to the boundary and to leave the others unchanged. In practice, due to hanging nodes, this may also lead to some refinement in the interior, however.

*Example* 3.6. ($L^2$-*norm error*) Finally, we consider again the $L^2$-norm error estimate but for a problem with strongly varying diffusion coefficient,

$$-\nabla \cdot \{a \nabla u\} = f \quad \text{in } \Omega, \quad u_{|\partial\Omega} = 0. \tag{3.16}$$

This example is intended to show that even when estimating the error in global norms, it can be beneficial to keep the dual weights within the error estimator rather than condensing them into just one global stability constant. In the considered situation the dual problem reads in strong form

$$-\nabla \cdot \{a \nabla z\} = e\|e\|^{-1} \quad \text{in } \Omega, \quad z_{|\partial\Omega} = 0, \tag{3.17}$$

and the local residuals take the form

$$R_{h|K} = f + \nabla \cdot \{a \nabla u_h\}, \quad r_{h|\Gamma} = \tfrac{1}{2} n \cdot [a \nabla u_h].$$

By the same argument which led us to Proposition 3.1 and then to the standard $L^2$-error estimate, we infer the following two types of a posteriori error estimates:

- 'Weighted' error estimate:

$$\|e\| \leq \eta_{L^2}^{\omega} := c_I \sum_{K \in \mathbb{T}_h} h_K^2 \rho_K \omega_K, \quad \omega_K := \|\nabla^2 z\|_K. \qquad (3.18)$$

- 'Global' error estimate:

$$\|e\| \leq \eta_{L^2} := c_I c_S \Big( \sum_{K \in \mathbb{T}_h} h_K^4 \, \rho_K^2 \Big)^{1/2}, \quad c_S := \|\nabla^2 z\|_{\Omega}. \qquad (3.19)$$

The residual terms $\rho_K$ are defined as before. In both cases, the stability terms can be evaluated by replacing the true dual solution by its Galerkin approximation $\|\nabla^2 z\|_K \approx \|\nabla_h^2 z_h\|_K$, with some second-order difference operator $\nabla_h^2$. The interpolation constant is typically of size $c_I \approx 0.2$. The error-dependent functional $J(\cdot) = (\cdot, e)\|e\|^{-1}$ is evaluated by replacing the unknown solution $u$ by a patch-wise higher-order interpolation $I_{2h}^{(2)} u_h$ of $u_h$,

$$e \approx I_{2h}^{(2)} u_h - u_h.$$

This gives us approximate $L^2$-error estimators denoted by $\tilde{\eta}_{L^2}^{\omega}(u_h)$ and $\tilde{\eta}_{L^2}(u_h)$, respectively. We want to compare the performance of these two $L^2$-error estimators by a numerical experiment. To this end, consider the particular setting $\Omega = (-1,1)^2$ and $a(x) = 0.1 + e^{3(x_1 + x_2)}$, with a sinusoidal solution $u(x)$ and corresponding right-hand side $f$. In this calculation the mesh adaptation tries to equilibrate the local 'error indicators' $\eta_K = h_K^2 \rho_K \omega_K$ and $\eta_K = h_K^4 \rho_K^2$, respectively. (This and alternative strategies for mesh adaptation will be discussed in more detail in the next chapter.)

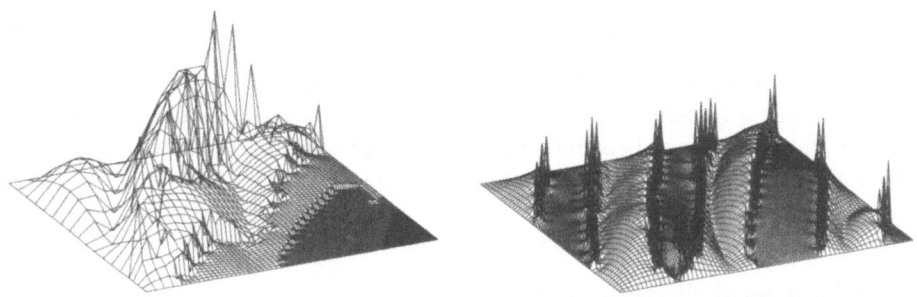

Figure 3.4: *Point-value errors obtained by $\eta_{L^2}$ (left) and $\eta_{L^2}^{\omega}$ (right, scaled by 1 : 3) on meshes with $N \approx 10,000$ cells; from Becker and Rannacher [30].*

Table 3.1: *Results obtained by $\tilde{\eta}_{L_2}$ and $\tilde{\eta}_{L_2}^\omega$; from Becker and Rannacher [30].*

| | $\tilde{\eta}_{L^2}$ | | | | $\tilde{\eta}_{L^2}^\omega$ | | |
|---|---|---|---|---|---|---|---|
| $TOL$ | $N$ | $\|e\|$ | $\tilde{\eta}_{L^2}$ | $\tilde{c}_s$ | $N$ | $\|e\|$ | $\tilde{\eta}_{L^2}^\omega$ |
| $4^{-2}$ | 2836 | $6.40 \cdot 10^{-2}$ | $2.32 \cdot 10^{-1}$ | 3.62 | 64 | $1.47 \cdot 10^{-1}$ | $1.52 \cdot 10^{-1}$ |
| $4^{-3}$ | 5884 | $2.13 \cdot 10^{-2}$ | $1.21 \cdot 10^{-1}$ | 5.68 | 148 | $1.08 \cdot 10^{-1}$ | $9.80 \cdot 10^{-2}$ |
| $4^{-4}$ | 15736 | $7.36 \cdot 10^{-3}$ | $4.76 \cdot 10^{-2}$ | 6.46 | 220 | $6.77 \cdot 10^{-2}$ | $5.24 \cdot 10^{-2}$ |
| $4^{-5}$ | 23380 | $5.59 \cdot 10^{-3}$ | $3.12 \cdot 10^{-2}$ | 5.58 | 592 | $2.21 \cdot 10^{-2}$ | $2.59 \cdot 10^{-2}$ |
| $4^{-6}$ | | | | | 892 | $1.19 \cdot 10^{-2}$ | $1.54 \cdot 10^{-2}$ |
| $4^{-7}$ | | | | | 2368 | $5.11 \cdot 10^{-3}$ | $7.17 \cdot 10^{-3}$ |
| $4^{-8}$ | | | | | 3640 | $2.53 \cdot 10^{-3}$ | $3.72 \cdot 10^{-3}$ |

Figure 3.4 shows the (scaled) error distribution on meshes obtained by the two estimators. The results shown in Table 3.1 indicate that efficient control of the $L^2$-norm error in the case of heterogeneous coefficients requires the use of 'weighted' a posteriori error estimates, i.e., the dual weights should be explicitly kept in the estimator and evaluated computationally.

*Remark* 3.4. (*Curved boundaries*) All examples considered so far had been posed on polygonal domains which can be exactly matched by the finite element mesh domain, i.e.,

$$\Omega_h := \cup\{K \in \mathbb{T}_h\} = \bar{\Omega}.$$

This assumption largely simplifies the error analysis and the resulting error estimators. However, in many practical cases at least parts of the boundary of the domain are curved and cannot be matched exactly by a polynomial approximation, e.g., in the cylinder-flow examples presented in the Introduction. Therefore, we have to deal with this complication.

Figure 3.5: *Standard situations of cells with curved edges.*

We consider the two typical situations depicted by Figure 3.5 in which a cell $K$ at the boundary has a curved edge $\Gamma_K \subset \partial\Omega$. The computational domain can be made to satisfy $\Omega_h = \bar{\Omega}$, by simply extending ot truncating the domain of definition of shape functions. Shape functions and transformations are left unchanged. This approximation results in a *non-conforming* finite element scheme,

$$(\nabla u_h, \nabla \psi_h) = (f, \psi_h) \quad \forall \psi_h \in V_h, \tag{3.20}$$

in which $V_h \not\subset V$, since the elements of $V_h$ will usually not satisfy zero boundary conditions. For the error analysis, we assume that the error functional $J(\cdot)$ has an $L^2$ representation, i.e., there is a $j \in L^2(\Omega)$, such that $J(\varphi) = (\varphi, j)$. Situations in which this is not the case require special considerations.

Using the solution $z \in V$ of the dual problem

$$-\Delta z = j \quad \text{in } \Omega, \quad z_{|\partial\Omega} = 0, \tag{3.21}$$

we obtain the error identity

$$J(e) = (j, e) = (e, -\Delta z) = (\nabla e, \nabla z) - (e, \partial_n z)_{\partial\Omega}.$$

For any $\psi_h \in V_h$, there holds

$$(\nabla e, \nabla \psi_h) = (\nabla u, \nabla \psi_h) - (\nabla u_h, \nabla \psi_h)$$
$$= (-\Delta u, \psi_h) + (\partial_n u, \psi_h)_{\partial\Omega} - (f, \psi_h) = (\partial_n u, \psi_h)_{\partial\Omega},$$

and consequently,

$$J(e) = (\nabla e, \nabla(z - \psi_h)) - (\partial_n u, z - \psi_h)_{\partial\Omega} + (u_h, \partial_n z)_{\partial\Omega}.$$

Now, integrating by parts on each cell $K \in \mathbb{T}_h$, we obtain

$$J(e) = \sum_{K \in \mathbb{T}_h} \left\{ (f + \Delta u_h, z - \psi_h)_K + (\partial_n e, z - \psi_h)_{\partial K} \right\}$$
$$- (\partial_n u, z - \psi_h)_{\partial\Omega} + (u_h, \partial_n z)_{\partial\Omega}.$$

This implies the error representation

$$J(e) = \sum_{K \in \mathbb{T}_h} \left\{ (R_h, z - \psi_h)_K + (r_h, z - \psi_h)_{\partial K} \right\} + (u_h, \partial_n z)_{\partial\Omega}, \tag{3.22}$$

with cell and edge residuals defined as above by $R_{h|K} := f + \Delta u_h$ and

$$r_{h|\Gamma} := \begin{cases} \frac{1}{2}[\partial_n u_h] & \text{if } \Gamma \subset \partial K \setminus \partial\Omega, \\ -\partial_n u_h & \text{if } \Gamma \subset \partial\Omega. \end{cases}$$

We note that the situation considered above is not very practical since the evaluation of the discrete equations (3.20) requires the use of numerical integration which in general introduces further errors that are not considered here.

*Remark* 3.5. (*Nonhomogeneous Dirichlet data*) Nonhomogeneous Dirichlet data

$$u = g \quad \text{on} \quad \partial\Omega,$$

are usually treated by introducing a representative function $\hat{u} \in H^1(\Omega)$ satisfying $\hat{u}_{|\partial\Omega} = g$, and a corresponding finite element approximation $\hat{u}_h$ which is the interpolation (or the $L^2$ projection) of $g$ along $\partial\Omega$. The solution is then sought as $u_h = \hat{u}_h + u_h^0$, where $u_h^0$ again has zero boundary values. Then, assuming again the domain to be polygonal or polyhedral, the error representation (3.8) takes the form

$$J(e) = \sum_{K \in \mathbb{T}_h} \left\{ (R_h, z - \psi_h)_K + (r_h, z - \psi_h)_{\partial K} \right\} - (g - g_h, \partial_n z)_{\partial\Omega}. \qquad (3.23)$$

*Remark* 3.6. (*Neumann boundary conditions*) The treatment of Neumann boundary conditions

$$\partial_n u = g \quad \text{on} \quad \Gamma_N \subset \partial\Omega,$$

does not cause any problems even in the case of a curved boundary, provided that the mesh $\mathbb{T}_h$ is compatible with the decomposition of the boundary $\partial\Omega = \Gamma_D \cup \Gamma_N$. In this case, we simply assume the cells adjacent to the Neumann boundary to have possibly a curved edge or face matching the boundary exactly. Now, the variational formulation reads

$$a(u, \psi) = (f, \psi) + (g, \psi)_{\partial\Omega_N} \quad \forall \psi \in V,$$

where $V := \{ v \in H^1(\Omega), v_{|\Gamma_D} = 0 \}$. For the error of the corresponding Galerkin approximation, again Galerkin orthogonality holds, i.e., $a(e, \psi_h) = 0$ for $\psi_h \in V_h \subset V$. Then, the error representation (3.8) remains valid with the only modification that the edge residuals are now defined by

$$r_{h|\Gamma} := \begin{cases} \frac{1}{2}[\partial_n u_h], & \text{if } \Gamma \subset \partial K \backslash \partial\Omega, \\ 0, & \text{if } \Gamma \subset \Gamma_D, \\ g - \partial_n u_h, & \text{if } \Gamma \subset \Gamma_N. \end{cases}$$

## 3.4 Higher-order finite elements

We briefly describe the use of the DWR method for higher-order finite elements and will see that it can be used in this case without essential changes. Let $V_h^{(p)} \subset V$, be finite element spaces of order $p+1$, i.e., they possess the local approximation properties of polynomials of degree $p$. We recall that by setting $\psi_h = I_h^{(p)} z$ in (3.8), we have:

$$J(e) = \sum_{K \in \mathbb{T}_h} \left\{ (R_h, z - I_h^{(p)} z)_K + (r_h, z - I_h^{(p)} z)_{\partial K} \right\},$$

with the cell- and edge-residuals $R_h$ and $r_h$ as defined above and some local interpolation $I_h^{(p)} z \in V_h^{(p)}$. In order to extract cell-error indicators for local mesh adaptation, we may proceed as before in the low-order case:

$$
\begin{aligned}
|J(e)| &\le \sum_{K \in \mathbb{T}_h} |(R_h, z - I_h^{(p)} z)_K + (r_h, z - I_h^{(p)} z)_{\partial K}| \\
&\le \sum_{K \in \mathbb{T}_h} \{ \|R_h\|_K \|z - I_h^{(p)} z\|_K + \|r_h\|_{\partial K} \|z - I_h^{(p)} z\|_{\partial K} \}.
\end{aligned}
$$

This does not reduce the asymptotic efficiency, as will be be demonstrated for the special case $p = 2$. First, we collect the following estimates for the cell residuals and weights which are obtained by using standard trace estimates:

$$
\begin{aligned}
\|R_h\|_K &= \|f + \Delta u_h\|_K = \|\Delta e\|_K \\
h_K^{-1/2} \|r_h\|_{\partial K} &= \tfrac{1}{2} h_K^{-1/2} \|[\partial_n e]\|_{\partial K} \\
&\le c h_K^{-1/2} \{ h_K^{-1/2} \|\nabla e\|_{\tilde{K}} + h_K^{1/2} \|\nabla^2 e\|'_{\tilde{K}} \} \\
&\le c h_K^{-1} \|\nabla e\|_{\tilde{K}} + \|\nabla^2 e\|'_{\tilde{K}},
\end{aligned}
$$

where the prime in $\| \cdot \|'_{\tilde{K}}$ refers to a cell-wise evaluation of the norm, and $\tilde{K}$ is a patch of cells around $K$. Further there holds the higher-order interpolation estimate

$$
\|z - I_h^{(p)} z\|_K + h_K^{1/2} \|z - I_h^{(p)} z\|_{\partial K} \le c h_K^k \|\nabla^k z\|_{\tilde{K}}, \quad k = 1, 2, 3.
$$

Using the above estimates, we obtain on a quasi-uniform mesh, i.e. for $h_K \approx h$, that

$$
|J(e)| \le c \{ h^{-1} \|\nabla e\| + \|\nabla^2 e\|' \} h^k \|\nabla^k z\| \le c h^{k+1} \|\nabla^3 u\| \|\nabla^k z\|, \quad k = 1, 2, 3.
$$

We evaluate this error estimate for the special output functionals

$$
J(\varphi) := (\nabla \varphi, \nabla e) \|\nabla e\|^{-1}, \quad J(\varphi) := (\varphi, e) \|e\|^{-1},
$$

and

$$
J_\psi(\varphi) := (\varphi, \psi) \|\nabla \psi\|^{-1}, \quad \psi \in H_0^1(\Omega),
$$

which correspond to the energy-norm error, the $L^2$-norm error, and the error in the $H^{-1}$ norm $\| \cdot \|_{-1}$, i.e. the norm of the dual space $H^{-1}(\Omega)$ of $H_0^1(\Omega)$. In virtue of the corresponding a priori bounds for the dual solution (for sufficiently regular domains $\Omega$), we obtain the $L^2$ and energy error estimate

$$
\|e\| + h \|\nabla e\| \le c h^3 \|\nabla^3 u\|, \tag{3.24}
$$

as well as the 'negative-norm' error estimate

$$
\|e\|_{-1} := \sup_{\psi \in H_0^1(\Omega)} \frac{(e, \psi)}{\|\nabla \psi\|} \le c h^4 \|\nabla^3 u\|, \tag{3.25}
$$

which are all of optimal order.

## 3.5   Exercises

*Exercise* 3.1. Functional-oriented a posteriori error estimates can also be stated in terms of energy-norm error bounds. Consider the Poisson model problem. With the primal and dual errors $e := u - u_h$ and $e^* := z - z_h$, respectively, there holds

$$|J(e)| = (\nabla e, \nabla z) = (\nabla e, \nabla e^*) \leq \|\nabla e\| \, \|\nabla e^*\|$$

Then, any a posteriori bound for the energy-norm error supplies also a bound for $J(e)$. Specify a situation in which this simple minded approach is inefficient. Why is this approach not suited to extract refinement indicators from the error estimate?

*Exercise* 3.2. Let $\Omega \subset \mathbb{R}^2$ be a convex polygonal domain. Develop a residual-based a posteriori estimate for the error $e := u - u_h$ with respect to the $L^\infty$-norm employing a *global* stability constant. Use the weighted a priori estimate

$$\|r \nabla^2 g_\varepsilon^a\| \leq c_\Omega \, |\log(\varepsilon)|,$$

for the regularized Green function $g_\varepsilon^a$.

*Exercise* 3.3 *(Practical exercise)*. Consider the Poisson problem

$$-\Delta u = 1 \text{ in } \Omega, \quad u_{|\partial\Omega} = 0,$$

on the domain $\Omega \subset (-1, 1)^2$, shown in the figure below. Compute the function values $u(a)$ and $\partial_1 u(a)$ at the point $a = (.75, .75)$.

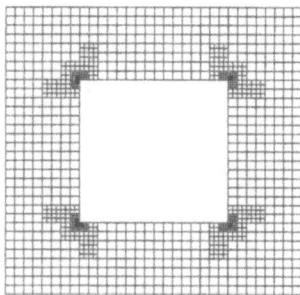

a) Use the provided program for computing $\partial_1 u(a)$ for a sequence of tolerances $\text{TOL}_i = 4^{-i}$, $i = 1, 2, \ldots$, and monitor the behavior of the true error and the number of cells depending on the achieved accuracy TOL. The code uses the duality-based weighted error estimator described in these Lecture Notes and the *fixed error fraction strategy* for mesh refinement.

b) Implement your own strategy of mesh refinement based on error indicators, smoothness indicators, or a priori information of your choice and try to beat the efficiency of the automatic error control process of (a).

# Chapter 4

# Practical Aspects

In this chapter, we discuss several aspects of the practical use of the DWR method described in the previous sections. These are (i) the practical and efficient evaluation of the a posteriori error representations, (ii) the extraction of local refinement indicators, and (iii) the design of strategies for economical mesh adaptation.

The starting point is the a posteriori *error representation*

$$J(e) = \sum_{K \in \mathbb{T}_h} \left\{ (R_h, z - \psi_h)_K + (r_h, z - \psi_h)_{\partial K} \right\} =: E(u_h). \tag{4.1}$$

as derived above for the Poisson problem. For its evaluation, due to Galerkin orthogonality, the subtraction of the arbitrary element $\psi_h \in V_h$ could be suppressed. From this, we extract local 'refinement indicators' called $\eta_K$ of the form

$$\eta_K := \left| (R_h, z - \psi_h)_K + (r_h, z - \psi_h)_{\partial K} \right|,$$

which are used to steer the mesh adaptation. Collecting these error indicators, we obtain an a posteriori *error estimator*, i.e. an upper bound for the error,

$$|J(e)| \leq \eta := \sum_{K \in \mathbb{T}_h} \eta_K. \tag{4.2}$$

We emphasize that in practice it does not make much sense to estimate further in the indicators $\eta_K$, since we would inevitably lose sharpness of the error bound.

For practical use of the error representation (4.1) and the error bound (4.2), we have to approximate the terms involving the unknown dual solution $z$ resulting in approximate error representations $\tilde{E}(u_h)$ and error indicators $\tilde{\eta}_K$. This may be expensive while the evaluation of the residuals $R_h$ and $r_h$ is usually cheap. We have to distinguish two related questions:

- Sharpness of the approximate error representations $\tilde{E}(u_h)$?

- Effectivity of the approximate local error indicators $\tilde{\eta}_K$ for mesh refinement?

For measuring the accuracy of the resulting error estimators, we will use the so-called *(reciprocal) effectivity index* defined by

$$I_{\text{eff}} := \left| \frac{\tilde{E}(u_h)}{J(e)} \right|,$$

which represents the degree of overestimation and should desirably be close to one.

*Remark* 4.1. Often the physical quantity to be computed can be expressed in different forms which coincide on the 'continuous' level but differ from each other for 'discrete' functions and may lead to more or less robust approximations. A typical example is the mean normal flux, and, more interesting, the drag and lift coefficients computed from solutions of the Navier-Stokes equations. Both can be expressed as surface or, after integration by parts, as volume integrals. If the desired output functional is not properly defined on the solution space $V$, such as point evaluation in two or more dimensions, it requires regularization,

$$J_\varepsilon(u) = J(u) + o(\varepsilon), \quad \varepsilon = TOL.$$

In the following, we will mostly suppress the index $\varepsilon$ indicating this regularization.

*Remark* 4.2. In general, the dual problem has to be solved numerically. In the case of a nonlinear problem the first step is linearization which will be discussed in Chapter 6. Then, this 'linearized' dual problem is approximately solved in some finite element space $\tilde{V}_h \subset V$,

$$\tilde{z}_h \in \tilde{V}_h : \quad a'(u_h)(\varphi_h, \tilde{z}_h) = J(\varphi_h) \quad \forall \varphi_h \in \tilde{V}_h,$$

where not necessarily $\tilde{V}_h = V_h$. This will be discussed in more detail below.

*Remark* 4.3. If on the basis of a numerical approximation to the dual solution $z$ an approximate error representation $\tilde{E}(u_h)$ has been generated, one may hope to obtain an improved approximation to the target quantity by setting

$$\tilde{J}(u_h) := J(u_h) + \tilde{E}(u_h) \approx J(u).$$

This 'post-processing' can significantly improve the accuracy in computing $J(u)$, but the resulting error can then no more be estimated on the basis of the available information.

## 4.1   Evaluation of the error identity and indicators

Consider the approximation of the model Poisson problem

$$-\Delta u = f \ \text{ in } \Omega, \quad u_{|\partial\Omega} = 0, \tag{4.3}$$

on a polygonal (or polyhedral) domain $\Omega \subset \mathbb{R}^d \,(d = 2\,\text{or}\,3)$ by piecewise bilinear (or trilinear) finite elements (in short '$Q_1$ elements'). For this prototypical case, we will compare several common strategies of evaluating the error representation (4.1) and the corresponding local error indicators. Notice that the approximation of $z$ in $E(u_h)$ simply by its Ritz projection $z_h \in V_h$ does not work since, by Galerkin orthogonality, it would result in the useless approximation $\tilde{E}(u_h) = 0$.

**Approximation by a higher-order method**

A first possibility is to solve dual problem by using *biquadratic* finite elements on the current mesh yielding an approximation $z_h^{(2)} \in V_h^{(2)}$ to $z$. This yields the approximate error representation

$$ E^{(1)}(u_h) := \sum_{K \in \mathbb{T}_h} \left\{ \left( R_h, z_h^{(2)} - I_h z_h^{(2)} \right)_K + \left( r_h, z_h^{(2)} - I_h z_h^{(2)} \right)_{\partial K} \right\}, $$

and the corresponding local error indicators

$$ \eta_K^{(1)} = \left| \left( R_h, z_h^{(2)} - I_h z_h^{(2)} \right)_K + \left( r_h, z_h^{(2)} - I_h z_h^{(2)} \right)_{\partial K} \right|. $$

For the special situation considered below (see Table 4.1), this approximation results in an asymptotically optimal effectivity index, $\lim_{TOL \to 0} I_{\text{eff}}^{(1)} = 1$. This observed behavior is supported by the theoretical analysis presented in the next section. However, approximating the dual problem by a higher-order method than used for $u_h$ seems not very attractive and can actually be avoided in most cases. A compromise would be to compute only the residual with respect to the coarse-grid space $V_{2h}^{(2)}$ (biquadratics on mesh $\mathbb{T}_{2h}$) of the bilinear Ritz projection $z_h \in V_h$ and to perform a few steps of defect correction using the system matrix of $V_h$ with appropriate blockwise preconditioning.

**Approximation by higher-order interpolation**

A simplification is achieved by patch-wise higher-order interpolation of the bilinear Ritz projection $z_h \in V_h$ of $z$. Here, we only consider *biquadratic* interpolation in 2-D. On square blocks of four neighboring cells the 9 nodal values of $z_h$ are used to define a biquadratic interpolation $I_{2h}^{(2)} z_h$. This is then used in the error representation instead of $z$, resulting in the approximate error representation

$$ E^{(2)}(u_h) := \sum_{K \in \mathbb{T}_h} \left\{ \left( R_h, I_{2h}^{(2)} z_h - z_h \right)_K + \left( r_h, I_{2h}^{(2)} z_h - z_h \right)_{\partial K} \right\}, $$

and the corresponding local error indicators

$$ \eta_K^{(2)} = \left| \left( R_h, I_{2h}^{(2)} z_h - z_h \right)_K + \left( r_h, I_{2h}^{(2)} z_h - z_h \right)_{\partial K} \right|. $$

For the special situation considered below (see Table 4.1), this approximation also results in an almost optimal effectivity index, $\underline{\lim}_{TOL \to 0} I_{\text{eff}}^{(2)} \sim 1$.

*Remark* 4.4. In the case of higher-order finite elements , with $p \geq 2$ , the evaluation of the error identity (4.1) may be done in a similar way as for $p = 1$ employing a patch-wise interpolation $I_{2h}^{(p')} z_h$ , with $p' > p$ , of the Ritz projection $z_h \in V_h^{(p)}$ . For example, in the case of biquadratic elements ($p = 2$), on a $2 \times 2$-cell patch we have 25 nodal values which can be used to construct an interpolation of degree $p = 4$ , i.e., in this case we would use $I_{2h}^{(p+2)} z_h$ .

### Approximation by difference quotients

The error representation (4.1) is estimated by

$$|E(u_h)| \leq \sum_{K \in \mathbb{T}_h} \rho_K \, \omega_K ,$$

with the notation of Proposition 3.1. Applying the usual cell-wise interpolation estimates, we have

$$\omega_K^2 = \|z - I_h z\|_K^2 + h_K \|z - I_h z\|_{\partial K}^2 \leq c_I^2 h_K^4 \|\nabla^2 z\|_K^2 ,$$

with an interpolation constant $c_I \sim 0.1 \ldots 1$. Now, the second derivatives $\nabla^2 z$ are replaced by suitable second-order difference quotients $\nabla_h^2 z_h$ of the Ritz projection $z_h$ of $z$ . This may be even more simplified to the cell-error indicators

$$\eta_K^{(3)} = c_I h_K^{3/2} \rho_K(u_h) \|[\partial_n z_h]\|_{\partial K} .$$

For the corresponding error estimator

$$E^{(3)}(u_h) := c_I \sum_{K \in \mathbb{T}_h} h_K^{3/2} \rho_K \|[\partial_n z_h]\|_{\partial K} ,$$

we usually observe strong over-estimation, i.e. $I_{\text{eff}}^{(3)} \gg 1$ (see Table 4.1) depending on what value we set for the interpolation constant $c_I$ .

### Approximation by local residual problems

On each cell $K$ , we solve the local Neumann problem (see Bank and Weiser [17])

$$(\nabla v_K, \nabla \psi_h)_K = (R_h, \psi_h)_K + (r_h, \psi_h)_{\partial K} \quad \forall \psi_h \in V_K ,$$

where $V_K = \{q \in \tilde{Q}_2(K),\ q \perp \tilde{Q}_1(K)\}$ . Then, in view of the relation

$$|(\nabla v_K, \nabla(z - I_h z))_K| \leq \|\nabla v_h\|_K \|\nabla(z - I_h z))_K\|$$
$$\leq c_I h_K \|\nabla v_h\|_K \|\nabla^2 z\|_K$$
$$\approx c_I h_K^{1/2} \|\nabla v_h\|_K \|[\partial_n z]\|_{\partial K} ,$$

the local error indicators may be defined as

$$\eta_K^{(4)} = c_I h_K^{1/2} \|\nabla v_K\|_K \|[\partial_n z_h]\|_{\partial K}.$$

In this way, one obtains fairly good bounds for the error in the energy norm. However, for local error functionals the results are not better than with the other simplified estimators. In such cases the crucial point seems more the accuracy in the approximation of the dual solution than in the evaluation of the residuals of $u_h$ (see Backes [13]).

## Numerical test

The effectivity of the first three of these error estimators has been tested for the 2-D model problem (4.3) with the solution $u(x) = (1 - x_1^2)(1 - x_2^2) \sin(4x_1) \sin(4x_2)$ for the two output functionals

$$J_1(u) := |S|^{-1} \int_S u \, dx, \quad S := [-\tfrac{1}{2}, 0] \times [0, \tfrac{1}{2}],$$

$$J_2(u) := u(\tfrac{1}{2}, \tfrac{1}{2}).$$

Table 4.1 shows the corresponding effectivity indices obtained on sequences of locally refined meshes on the basis of the local error indicators $\eta_K^{(i)}$, $i = 1, 2, 3$. It turns out that the cheap interpolation-based estimator $E^{(2)}(u_h)$ is almost as effective as the more expensive estimator $E^{(1)}(u_h)$. Therefore, in the following, we will almost exclusively use the first one in the presented numerical tests.

Table 4.1: *Effectivity of weighted error indicators for the mean error $J_1(e)$ (left) and the point-error $J_2(e)$ (right); from Richter [123].*

| $N$ | $J_1(e)$ | $I_{\text{eff}}^{(1)}$ | $I_{\text{eff}}^{(2)}$ | $I_{\text{eff}}^{(3)}$ | $N$ | $J_2(e)$ | $I_{\text{eff}}^{(1)}$ | $I_{\text{eff}}^{(2)}$ | $I_{\text{eff}}^{(3)}$ |
|---|---|---|---|---|---|---|---|---|---|
| 81 | $7.6 \cdot 10^{-2}$ | 1.01 | 1.05 | 393 | 81 | $4.2 \cdot 10^{-1}$ | 0.17 | 0.18 | 69 |
| 151 | $2.7 \cdot 10^{-2}$ | 1.00 | 1.07 | 337 | 151 | $1.4 \cdot 10^{-1}$ | 0.26 | 0.20 | 83 |
| 653 | $3.6 \cdot 10^{-3}$ | 0.99 | 0.99 | 229 | 635 | $1.0 \cdot 10^{-2}$ | 0.30 | 0.28 | 123 |
| 1435 | $1.4 \cdot 10^{-3}$ | 1.00 | 1.00 | 177 | 1443 | $2.9 \cdot 10^{-3}$ | 0.39 | 0.40 | 129 |
| 2937 | $6.5 \cdot 10^{-4}$ | 1.00 | 0.98 | 120 | 2875 | $9.4 \cdot 10^{-4}$ | 0.52 | 0.51 | 130 |
| 6249 | $3.2 \cdot 10^{-4}$ | 1.00 | 1.00 | 84 | 6229 | $2.8 \cdot 10^{-4}$ | 0.61 | 0.60 | 147 |
| 12995 | $1.2 \cdot 10^{-4}$ | 1.00 | 0.99 | 84 | 12521 | $1.0 \cdot 10^{-4}$ | 0.74 | 0.72 | 148 |
| 26603 | $7.2 \cdot 10^{-5}$ | 1.00 | 1.00 | 55 | 26903 | $3.9 \cdot 10^{-5}$ | 0.82 | 0.80 | 155 |

*Remark 4.5.* The traditional energy-norm error estimator $\eta_E$ is known to be not only *reliable*, i.e., it provides a safe upper bound for the energy error, but also *efficient*, i.e., it is asymptotically sharp in the sense that

$$c_1 \|\nabla e\| \le \eta_E \le c_2 \{\|\nabla e\| + \|f - \bar{f}_h\|\},$$

where $\bar{f}_h$ is the piecewise constant interpolation of $f$ (see, e.g., Verfürth [132]). This means that the estimator is up to a constant asymptotically correct. A corresponding result is not possible in general for estimators $\eta_\omega$ of locally defined error quantities. Already the transition from the *error representation* to the *error estimate* in terms of (non-negative) cell-wise error indicators is critical, since by this localization the asymptotic sharpness of the global error representation may get lost. To illustrate this, consider the case $J(u) = u(0)$ and assume that the exact as well as the approximate solution are anti-symmetric with respect to the $x_1$-axis. Then, $e(0) = 0$, but usually $\sum_{K \in \mathbb{T}_h} \eta_K \neq 0$.

## 4.2 Mesh adaptation

Next, we address the practical aspects of successive mesh adaptation. Suppose that on the meshes $\mathbb{T}_h$, we have local error indicators $\eta_K$ extracted from an a posteriori error estimate

$$|J(e)| \leq \eta := \sum_{K \in \mathbb{T}_h} \eta_K, \qquad N := \#\{K \in \mathbb{T}_h\}. \qquad (4.4)$$

Using this information the computational mesh may be adapted using various different strategies. For quadrilateral meshes, as considered here, the refinement and coarsening is facilitated by using *hanging nodes*. The global conformity of the finite element ansatz is preserved since the unknowns at hanging nodes are eliminated by interpolation between the neighboring 'regular' nodes.

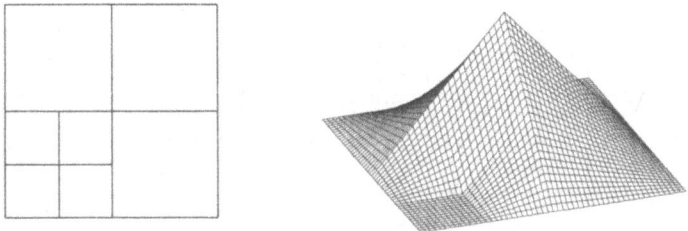

Figure 4.1: $Q_1$ *nodal basis function on a patch of cells with hanging nodes*

We note that there are several alternative strategies for realizing local mesh refinement. The occurrence of hanging nodes can be avoided by using special 'transition cells' (triangles or quadrilaterals) which bridge from cells of width $h$ to those of width $h/2$. The construction of such cells may be complicated in 3-D. It may also cause a spreading of the refinement zone which implies extra work for de-refining. However, it is basically a question of taste which technique of mesh organization one prefers. Refinement and coarsening of quadrilateral meshes involving the use of hanging nodes proceeds as indicated in Figure 4.2.

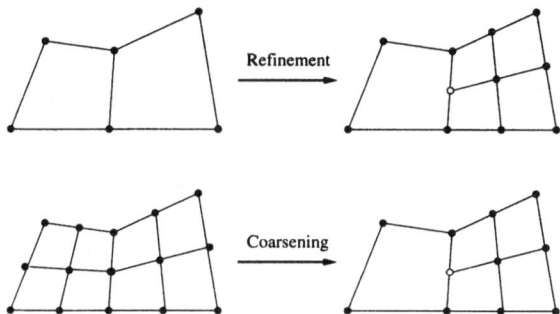

Figure 4.2: *Refinement and coarsening in quadrilateral meshes.*

At first, we have to check whether on the current mesh $\mathbb{T}_h$ the *stopping criterion*

$$\eta \leq TOL$$

is already satisfied. If this is the case, then $u_h$ is accepted as approximation to $u$ that represents the target quantity $J(u)$ by $J(u_h)$ within the desired tolerance $TOL$. Otherwise, the next refinement cycle is started. To this end, the cells in the current mesh are ordered according to

$$\eta_{K_1} \geq \cdots \geq \eta_{K_i} \geq \cdots \geq \eta_{K_N}.$$

Then, the mesh adaptation may be organized using one of the following strategies.

**Error-balancing strategy**

Below, we will give an argument which indicates that an *optimal* mesh is characterized by equilibrated error indicators, such as

$$\eta_{K_i} \approx \frac{TOL}{N}, \quad i = 1, \ldots N,$$

which implies $\eta(u_h) \approx TOL$. However, the so-called *error-balancing strategy* based on this idea is 'implicit' as it involves the current number of mesh cells $N$ which is obtained only at the end of the adaptation cycle. We check, starting from $i = 1, j = 0$, whether

$$\eta_{K_i} \leq \frac{TOL}{N + 3j}.$$

If this is not satisfied, then the cell $K_i$ is refined, the counters $j$ and $i$ are increased by one, and one proceeds to the next smaller $\eta_{K_i}$. But, if the condition is satisfied, then the new mesh $\mathbb{T}_h^{new}$ is reached. This strategy is potentially optimal but involves many expensive checking operations and is therefore impracticable.

**Fixed-error-reduction or fixed-rate strategy**

For fractions $X, Y$, with $1 - X > Y$, determine indices $N_*, N^* \in \{1, \dots, N\}$, such that

$$\sum_{i=1}^{N^*} \eta_{K_i} \approx X\eta, \qquad \sum_{i=N_*}^{N} \eta_{K_i} \approx Y\eta.$$

Then, the cells $K_1, \dots, K_{N^*}$ are refined and the cells $K_{N^*}, \dots, K_N$ are coarsened. Common choices are $X = 0.2$ and $Y = 0.1$. Alternatively, in the *fixed-rate strategy*, one refines $X \cdot N$ and coarsens $Y \cdot N$ cells with largest and smallest error indicators, respectively. For appropriate choices of $X, Y$ this accomplishes to keep the number of cells almost constant in the course of the mesh adaptation process.

**Mesh-optimization strategy**

The information contained in the error representation (4.1) may be used directly to construct an 'optimal' mesh on which the error tolerance $\eta \approx$ TOL is achieved, i.e. skipping the intermediate 'one-level' refinement steps. Here, a mesh $\mathbb{T}_h = \{K\}$ is characterized by a continuous *mesh-size function* $h(x)$, where $h_{|K} \approx h_K$. Further, it is assumed that the error estimator is related to a continuous limit of the form

$$\sum_{K \in \mathbb{T}_h} \eta_K \approx \int_\Omega h(x)^2 \Phi(x)\, dx =: \eta(h),$$

with $\Phi := (\Phi_1 + \Phi_2)\Phi_3$ a mesh-independent weighting function. We note that the latter requirement rules out cases in which the functional $J(\cdot)$ and hence also the dual solution implicitly depend on the mesh size, as for example in the estimation of the error in the energy or the $L^2$ norm (cf. Section 3.2). The components of $\Phi$ are defined by the limiting processes of residuals and weights, for TOL $\to 0$:

$$\max_{x \in K} |f + \Delta u_h| \approx \max_{x \in K} \Phi_1, \qquad \tfrac{1}{2} h_K^{-1} \max_{x \in K} |[\partial_n u_h]_{\partial K}| \approx \max_{x \in K} \Phi_2, \qquad (4.5)$$

$$h_K^{-2} \max_{x \in K} |z - I_h z| \approx \max_{x \in K} \Phi_3. \qquad (4.6)$$

Here, the mesh-size power $h(x)^2$ is related to the 'order' of the considered finite element ansatz. The justification of these assumptions will be discussed in the next chapter. The function $\Phi(x)$ may have strong (regularized) singularities which will possibly require the mesh-size $h(x)$ to reduce down towards zero even for tolerance $TOL > 0$. Further, we introduce the mesh complexity formula

$$N = \sum_{K \in \mathbb{T}_h} h_K^d h_K^{-d} \approx \int_\Omega h(x)^{-d}\, dx =: N(h),$$

which denotes the number of degrees of freedom for a mesh-size function $h(x)$. With this notation, we obtain the following result.

**Proposition 4.6.** *The mesh-optimization problem*

$$\eta(h) \to \min, \quad N(h) \leq N_{\max} \tag{4.7}$$

*is solved by*

$$h_{\text{opt}}(x) = \left(\frac{W}{N_{\max}}\right)^{1/d} \Phi(x)^{-1/(2+d)}, \tag{4.8}$$

*provided that*

$$W := \int_\Omega \Phi(x)^{d/(2+d)} \, dx < \infty.$$

*For an 'optimal' mesh, there hold*

$$TOL = \frac{W^{(2+d)/d}}{N^{2/d}} \quad and \quad N \approx \frac{W^{(2+d)/2}}{TOL^{d/2}}. \tag{4.9}$$

*Proof.* Following the classical Lagrange approach, we introduce the Lagrangian

$$L(h, \lambda) = \eta(h) + \lambda\{N(h) - N_{\max}\}$$

with Lagrangian multiplier $\lambda \in \mathbb{R}$. Then, the optimal mesh-size function $h_{\text{opt}}$ is characterized as stationary point of the first-order optimality condition

$$\frac{d}{dt}L(h+t\varphi, \lambda)_{|t=0} = 0, \quad \frac{d}{dt}L(h, \lambda+t\mu)_{|t=0} = 0,$$

for all admissible variations $\varphi$ and $\mu$. This means that

$$2h(x)\Phi(x) - d\lambda h(x)^{-d-1} = 0, \quad \int_\Omega h(x)^{-d} \, dx - N_{\max} = 0,$$

and, consequently,

$$h(x) = \left(\frac{2}{d\lambda}\Phi(x)\right)^{-1/(2+d)}, \quad \left(\frac{2}{d\lambda}\right)^{d/(2+d)} \int_\Omega \Phi(x)^{d/(2+d)} \, dx = N_{\max}.$$

From this, we deduce the desired relations

$$\lambda \equiv \frac{2}{d} h(x)^{2+d}\Phi(x), \quad h_{\text{opt}}(x) = \left(\frac{W}{N_{\max}}\right)^{1/d}\Phi(x)^{-1/(2+d)}.$$

From the formula for $h_{\text{opt}}$, we conclude that on the optimal mesh, there holds

$$TOL = \left(\frac{W}{N}\right)^{2/d} \int_\Omega \Phi(x)^{-2/(2+d)}\Phi(x) \, dx = \frac{W^{(2+d)/d}}{N^{2/d}},$$

which proves (4.9). $\qquad\square$

*Remark* 4.7. We note that in two dimensions the optimal mesh complexity is

$$TOL = \mathcal{O}(N^{-1}) \quad \text{or} \quad N = \mathcal{O}(TOL^{-1}),$$

for linear or bilinear finite elements, provided that $\sup_{TOL \to 0} W < \infty$ is satisfied. This is the case even for rather 'irregular' functionals $J(\cdot)$. For example, the evaluation of $J(u) = \partial_i u(a)$ leads to $\Phi(x) \approx (|x-a|^2 + TOL^2)^{-3}$ and, consequently,

$$\sup_{TOL \to 0} W \approx \int_{\Omega} |x-a|^{-3/2} \, dx < \infty.$$

In the case that $\sup_{TOL} W = \infty$, as for example for $J(u) = \partial_i^2 u(a)$, the optimal mesh complexity becomes

$$TOL = \mathcal{O}(N^{\alpha-1}) \quad \text{or} \quad N = \mathcal{O}(TOL^{-\alpha-1}),$$

with some $\alpha > 0$. In particular, for the latter functional, we easily find $TOL = \mathcal{O}(N^{-1}|\log(N)|)$.

*Remark* 4.8. The result of Proposition 4.6 implies that the optimal mesh-size distribution is characterized by the *equilibration property*

$$h(x)^{2+d}\Phi(x) \equiv \left(\frac{TOL}{W}\right)^{(2+d)/d} = \text{const.} \tag{4.10}$$

This justifies the strategy of equilibrating the local error indicators $\eta_K$,

$$\eta(h) = \int_{\Omega} h(x)^2 \Phi(x) \, dx \approx \sum_{K \in \mathbb{T}_h} h_K^{2+d} \Phi_K = \sum_{K \in \mathbb{T}_h} \eta_K,$$

as used in the *error-balancing strategy*.

Let the weight function $\Phi(x)$ be bounded. Then, once a balanced mesh satisfying (4.10) is reached, a maximum increase of accuracy is achieved by uniform mesh refinement. If $\Phi$ is singular, then the 'optimal' mesh size tends to zero at these singularities even for $TOL > 0$.

*Remark* 4.9. Alternatively to (4.7), we can also consider the mesh-optimization problem

$$N(h) \to \min, \quad \eta(h) \leq TOL,$$

which has the solution

$$h_{\text{opt}}(x) = \left(\frac{TOL}{W}\right)^{1/2} \Phi(x)^{-1/(2+d)}.$$

*Remark* 4.10. Although the mesh-optimization strategy seems very attractive, its realization involves several problems:

- The derivation of the formula for an optimal mesh-size distribution is based on the assumption that on the considered meshes the cell-residuals behave in an optimal way under refinement, i.e., $\rho_K \approx h_K$. This is hard to prove and may not be true in general; see the discussion of this question in Chapter 5.

- The numerical approximation of the weighting function $\Phi(x)$ should provide more information than can be cheaply obtained using only information on the current mesh.

- The explicit formulas for $h_{\text{opt}}(x)$ have to be used with care in designing a mesh as their derivation implicitly assumes that they actually correspond to *scalar* mesh-size functions of *isotropic* meshes, a condition, however, which is not incorporated into the formulation of the mesh-optimization problems.

A detailed study of how to utilize mesh optimization for the model problem has been made by Richter [123].

*Remark* 4.11. *(The role of regularization)* At the beginning of this chapter, we had mentioned that in the case of a 'singular' functional like, for example, that for evaluation of the derivative point value, regularization is necessary,

$$J_\varepsilon(u) := |B_\varepsilon|^{-1} \int_{B_\varepsilon} \partial_1 u \, dx,$$

To demonstrate the effect caused by regularizing with $\varepsilon = TOL$, as proposed above, compared to $\varepsilon = h_{min}$, we show the results of a numerical test for the model situation considered in Example 3.4, see Table 4.2. We see that in this case regularization drastically reduces the number $L$ of refinement steps for reaching the same accuracy level which means less computational work.

Table 4.2: *Results for computing $\partial_1 u(0)$ in Example 3.4 using regularization with $\varepsilon = TOL$ (right) and $\varepsilon = h_{min}$ (left); from Becker and Rannacher [30].*

| TOL | $\varepsilon = h_{min}$ | | | $\varepsilon = TOL$ | | |
|---|---|---|---|---|---|---|
| | N | L | $|\partial_1 e(0)|$ | N | L | $|\partial_1 e(0)|$ |
| 1 | 40 | 4 | $2.57 \cdot 10^{-0}$ | 40 | 4 | $2.57 \cdot 10^{-0}$ |
| $4^{-1}$ | 124 | 6 | $7.38 \cdot 10^{-1}$ | 64 | 4 | $1.47 \cdot 10^{-0}$ |
| $4^{-2}$ | 448 | 11 | $1.35 \cdot 10^{-3}$ | 148 | 6 | $7.51 \cdot 10^{-1}$ |
| $4^{-3}$ | 1780 | 15 | $1.19 \cdot 10^{-3}$ | 940 | 9 | $4.10 \cdot 10^{-1}$ |
| $4^{-4}$ | 6328 | 19 | $2.20 \cdot 10^{-3}$ | 4912 | 12 | $4.14 \cdot 10^{-3}$ |
| $4^{-5}$ | 25984 | 24 | $4.28 \cdot 10^{-4}$ | 20980 | 15 | $2.27 \cdot 10^{-4}$ |
| $4^{-6}$ | 95260 | 28 | $1.39 \cdot 10^{-4}$ | 86740 | 17 | $5.82 \cdot 10^{-5}$ |

## 4.3   Use of error estimators for post-processing

The duality approach described so far for estimating the discretization error can also be used for increasing the accuracy in the approximation of the target quantity. Consider the model situation as before, i.e. the Poisson problem in 2-D written in variational form as

$$a(u, \psi) = (f, \psi) \quad \psi \in V, \tag{4.11}$$

and discretized by

$$a(u_h, \psi_h) = (f, \psi_h) \quad \psi_h \in V_h. \tag{4.12}$$

The output functional is $J(\cdot)$. Let $z \in V$ be the corresponding dual solution and $z_h \in V_h$ its Ritz projection on the 'primal' mesh $\mathbb{T}_h$. By definition, there holds

$$J(u) = a(u, z) = (f, z), \tag{4.13}$$

i.e., if $z$ were known, the target quantity $J(u)$ is determined solely by the data $f$. This observation will be used below for post-processing the approximation $J(u_h)$. For this, we recall the identity

$$J(u) = J(u_h) + a(e, z) = J(u_h) + \rho(u_h)(z - z_h),$$

where

$$\rho(u_h)(z - z_h) = (f, z - z_h) - a(u_h, z - z_h).$$

Above, we have discussed ways of constructing approximations to the dual solution $z$, e.g. the patch-wise biquadratic interpolation $\tilde{z}_h := I_{2h}^{(2)} z_h$ of $z_h$ on the mesh $\mathbb{T}_h$. This led us to the approximate error representation

$$J(e) \approx \rho(u_h)(\tilde{z}_h - z_h) = \rho(u_h)(\tilde{z}_h).$$

Rewriting this relation as

$$J(u) \approx \tilde{J}_1(u_h) := J(u_h) + (f, \tilde{z}_h) - a(u_h, \tilde{z}_h),$$

we obtain a presumably better approximation to $J(u)$ than is $J(u_h)$. The error can be written as

$$J(u) - \tilde{J}_1(u_h) = J(e) - \rho(u_h)(\tilde{z}_h) = a(e, z) - a(u_h, \tilde{z}_h) = a(e, z - \tilde{z}_h),$$

which implies

$$|J(u) - \tilde{J}_1(u_h)| \leq \|\nabla e\| \, \|\nabla(z - \tilde{z}_h)\|. \tag{4.14}$$

Since it is not clear whether $\tilde{z}_h$ is a reasonably better approximation to $z$ than $z_h$, this estimate is of only questionable value. Further, the two energy-norm errors correspond both to the 'primal' mesh $\mathbb{T}_h$ and can therefore not be minimized independently. It would be desirable to have the possibility of using independent meshes $\mathbb{T}_h$ for $u_h$ and $\mathbb{T}_h^*$ for $z_h$ in constructing an approximation of $J(u)$. This is accomplished in the following proposition (see Richter [123]).

**Proposition 4.12.** *Let* $\mathbb{T}_h$ *and* $\mathbb{T}_h^*$ *be two independent meshes and* $V_h$ *and* $V_h^*$ *corresponding finite element spaces in which the Ritz projections* $u_h$ *and* $z_h^*$ *of* $u$ *and* $z$ *are computed. Further, denote by* $\tilde{z}_h^*$ *the patch-wise biquadratic interpolation of* $z_h^*$ *on the dual mesh* $\mathbb{T}_h^*$ *. Then, for the post-processed approximation*

$$\tilde{J}_2(u_h) := J(u_h) + (f, \tilde{z}_h^*) - a(u_h, \tilde{z}_h^*),$$

*there holds the estimate*

$$|J(u) - \tilde{J}_2(u_h)| \leq \|\nabla e\| \, \|\nabla(z - \tilde{z}_h^*)\|. \tag{4.15}$$

*Proof.* We have

$$
\begin{aligned}
J(u) - \tilde{J}_2(u_h) &= J(u) - J(u_h) - \rho(u_h)(\tilde{z}_h^*) \\
&= (f, z) - a(u_h, z) - (f, \tilde{z}_h^*) + a(u_h, \tilde{z}_h^*) \\
&= (f, z - \tilde{z}_h^*) - a(u_h, z - \tilde{z}_h^*) \\
&= a(e, z - \tilde{z}_h^*).
\end{aligned}
$$

This implies the assertion. □

We emphasize that in the estimate (4.15) the two energy-norm terms can be minimized independently by optimizing the primal and dual meshes $\mathbb{T}_h$ and $\mathbb{T}_h^*$, respectively. One may hope to obtain an even better approximation by

$$\tilde{J}_3(u_h) := \tilde{J}_2(\tilde{u}_h) = J(\tilde{u}_h) + (f, \tilde{z}_h^*) - a(\tilde{u}_h, \tilde{z}_h^*),$$

where $\tilde{u}_h$ is the patch-wise biquadratic interpolation of $u_h$. Below, we will test all three post-processed approximations to $J(u)$ for the model problem on $\Omega = (-1, 1)^2$ with the prescribed solution $u(x) = (1 - x_1^2)(1 - x_2^2) \exp(1 - x_2^{-4})$ and the output functional

$$J(u) = \int_{-1}^{1} u(x_1, 0.5) \, dx.$$

In this example primal and dual solution have irregularities at different locations such that it is to expected that maximal efficiency is achieved using different meshes for $u_h$ and $z_h$ as shown in Figure 4.3.

Figure 4.4 shows the mesh efficiencies of $J(u_h)$ and the three post-processed approximations $\tilde{J}_1(u_h)$, $\tilde{J}_2(u_h)$, and $\tilde{J}_3(u_h)$. The mesh refinements are driven by energy-norm error indicators as derived in Section 3.2 separately on the primal and dual meshes $\mathbb{T}_h$ and $\mathbb{T}_h^*$, respectively. We see that $\tilde{J}_1(u_h)$ indeed does not bring significant advantages over the original approximation $J(u_h)$. The two other approximations $\tilde{J}_2(u_h)$ and $\tilde{J}_3(u_h)$ which use different meshes for $u$ and $z$ are clearly superior and show a mesh complexity like $TOL \approx N^{-2}$ ($N = N_{\text{primal}} + N_{\text{dual}}$). This is to be compared to the optimal complexity $TOL \approx N^{-1}$ obtained above in Proposition 4.12.

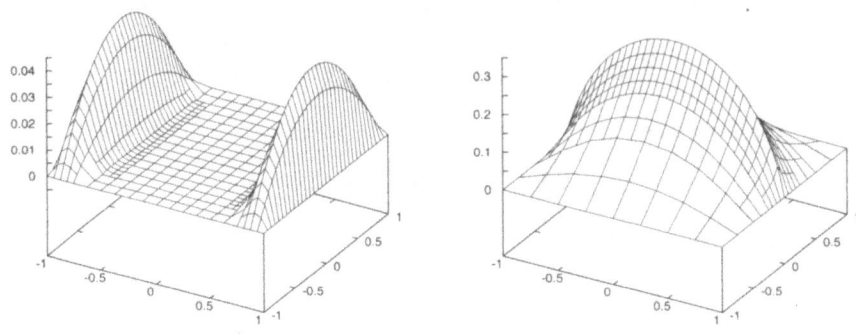

Figure 4.3: *Primal (left) and dual (right) solution of the model problem; from Richter [123].*

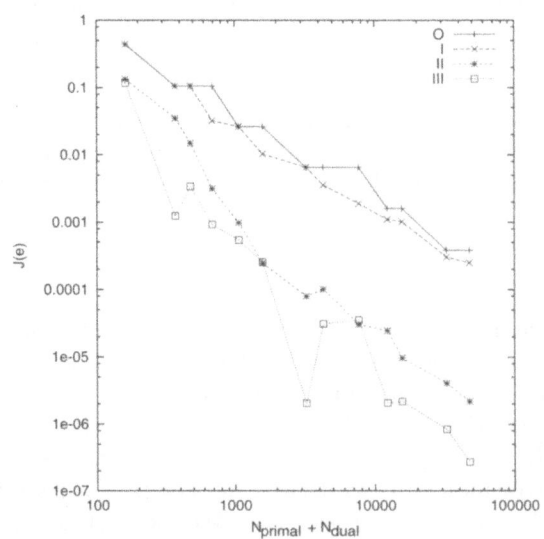

Figure 4.4: *Mesh efficiencies of post-processing: $J(u_h)$ (solid line, symbol $+$), $\tilde{J}_1(u_h)$ (broken line, symbol $\times$), $\tilde{J}_2(u_h)$ (broken line, symbol $*$), $\tilde{J}_3(u_h)$ (dotted line, symbol $\square$); from Richter [123].*

## 4.4 Towards anisotropic mesh adaptation

Sometimes *isotropic* mesh refinement as discussed so far is not efficient for properly resolving direction-dependent features of the solution. For example, in singularly perturbed problems of the form

$$-\varepsilon \Delta u + bu = f \text{ in } \Omega, \qquad u_{|\partial\Omega} = 0,$$

with small coefficient $\varepsilon$, boundary layers may occur in which the solution has a large derivative in normal direction to the boundary while it varies only slowly in tangential direction. A similar phenomenon occurs in 3D along reentrant edges of the domain (edge singularities). In such a situation it is appropriate to use meshes which are anisotropically refined in the sense that the cells along the boundary are much thinner in normal than in tangential direction, see Figure 4.5.

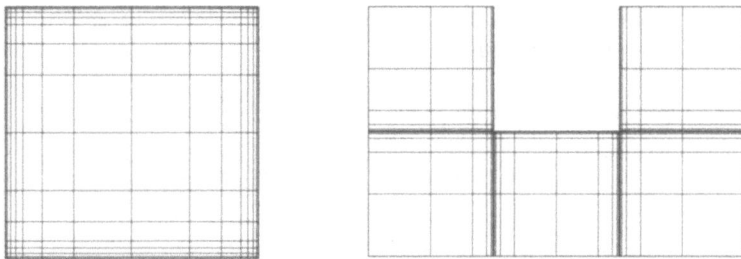

Figure 4.5: *Locally anisotropic tensor-product meshes.*

The questions are now whether the weighted error estimator contains information about anisotropy in the exact solution, and whether we can extract local indicators which tell us how to adapt the mesh according to (i) orientation of cells, and (ii) optimal stretching of cells. This aspect of automatic mesh adaptation is a rather difficult one and still the subject of current research.

To fix ideas, consider a given function $u \in C^2(\bar{B}_1)$ on the ball $B_1 = \{x \in \mathbb{R}^d, |x| < 1\}$ and determine the direction of smallest line-wise $L^2$ error of linear interpolation, i.e., find the unit vector $e$ such that

$$\int_\Gamma |u - I_h u|^2 \, ds \approx \int_\Gamma |\partial_e^2 u|^2 \, ds = \int_\Gamma |(H(u)e, e)| \, ds \; \longrightarrow \; \min_\Gamma .$$

Here, $\Gamma = \{x \in B_1, x = se, 0 \le s < 1\}$, and $H(u) := \nabla^2 u$ is the (symmetric) Hessian matrix of $u$. Hence, the interpolation error becomes minimal for $e = e_{\min}$ being the eigenvector corresponding to the eigenvalue of $H(u)$ with minimal modulus. We see that all the information about the best orientation of cells for minimizing the interpolation error is available by the Hessian matrix. The eigenvalue quotient $\sigma = |\lambda_{\max}/\lambda_{\min}|$ determines the cell aspect ratio and the corresponding (orthogonal) eigenvectors $e_{\min}$ and $e_{\max}$ the orientation of the cell.

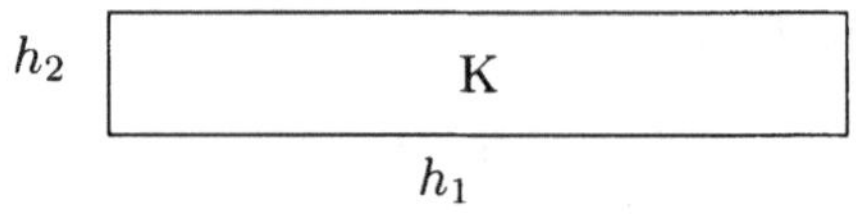

Figure 4.6: *Cell of a Cartesian mesh.*

Quadrilateral meshes are not very suited for dynamical cell reorientation. For this purpose, triangular meshes are more flexible. Therefore, we will consider adaptive cell stretching alone and, for simplicity, will concentrate on the construction of 'optimal' Cartesian meshes consisting of cells $K$ with area $|K| = h_1 h_2$, as shown in Figure 4.6. Again 'hanging' nodes are allowed. The maximum $\sigma_K := \max\{h_1/h_2, h_2/h_1\}$ is the *cell aspect ratio* and $\sigma_h := \max_{K \in \mathbb{T}_h} \sigma_K$ the *maximum mesh aspect ratio*.

We recall the a posteriori error representation

$$J(e) = \sum_{K \in \mathbb{T}_h} \left\{ (f + \Delta u_h, z - I_h z)_K + \tfrac{1}{2}([\partial_n u_h], z - I_h z)_{\partial K \setminus \partial \Omega} \right\}. \quad (4.16)$$

Practical experience and theoretical analysis show that on *isotropic* meshes the edge-residual terms $([\partial_n u_h], z - I_h z)_{\partial K}$ can be made to dominate the cell-residual terms $(f + \Delta u_h, z - I_h z)_K$, see Exercise 5.2. Let us assume that the same is true also for anisotropic meshes (for some theoretical support see Kunert and Verfürth [99]). Therefore, we only consider the edge-residual terms and as before suppose the approximation

$$h_i^{-1}[\partial_i u_h] \approx \partial_i^2 u \quad (i = 1, \ldots, d).$$

Then, assuming again the second-order derivatives of $u$ as constant, we obtain

$$|([\partial_n u_h], z - I_h z)_{\partial K}| \approx h_1^3 h_2 |\partial_2^2 u| \, |\partial_1^2 z| + h_1 h_2^3 |\partial_1^2 u| \, |\partial_2^2 z|$$
$$= |K|\{h_1^2 |\partial_2^2 u| \, |\partial_1^2 z| + |K|^2 h_1^{-2} |\partial_1^2 u| \, |\partial_2^2 z|\}.$$

Minimizing this with respect to $h_1$ yields the necessary condition

$$2h_1 |\partial_2^2 u| \, |\partial_1^2 z| - 2|K|^2 h_1^{-3} |\partial_1^2 u| \, |\partial_2^2 z| = 0 \quad \Rightarrow \quad h_1^4 = |K|^2 \frac{|\partial_1^2 u| \, |\partial_2^2 z|}{|\partial_2^2 u| \, |\partial_1^2 z|},$$

and, consequently,

$$\frac{h_1}{h_2} \approx \left( \frac{|\partial_1^2 u| \, |\partial_2^2 z|}{|\partial_2^2 u| \, |\partial_1^2 z|} \right)^{1/2}. \quad (4.17)$$

This result is counterintuitive as it does not indicate the optimal cell stretching. To see this, consider the case that $u$ is linear in $x_1$-direction, i.e., $\partial_1^2 u \equiv 0$, and that $z$ is isotropic. Then, formula (4.17) would suggest to refine the cell in $x_1$ direction which is evidently the wrong decision. It seems that considering only the edge terms in (4.16) leads to contradictory results, and we rather have to take into account the whole combination of cell and edge residuals.

*Remark* 4.13. The development of a rigorous criterion for anisotropic mesh refinement on the basis of 'goal-oriented' error representations such as (4.16) must be left as an open problem. Here the difficulty is caused by the local interplay of the primal and dual solutions which may have significanty different regularity properties. In the special case of error estimation with respect to the energy-norm the dual solution coincides with the primal error such that both quantities behave very much the same way. Then, anisotropic mesh refinement can be surely guided by information from the jump residuals of $u_h$ alone (see, e.g., Siebert [126]).

In view of the above discussion, we now follow a more heuristic approach and base the anisotropic cell adaptation on an estimate for the interpolation error. We recall the anisotropic interpolation error (see, e.g., Becker [18])

$$\|\nabla(u - I_h u)\|_K \le c \left( h_1^2 \|\partial_1 \nabla u\|_K^2 + h_2^2 \|\partial_2 \nabla u\|_K^2 \right)^{1/2}. \tag{4.18}$$

Hence, assuming the second-order derivatives as constant on $K$, we have

$$\|\nabla(u - I_h u)\|_K \le c |K|^{1/2} \left( h_1^2 |\partial_1 \nabla u|^2 + |K|^2 h_1^{-2} |\partial_2 \nabla u|^2 \right)^{1/2}.$$

Minimizing this with respect to $h_1$ results in the necessary condition

$$2h_1 |\partial_1 \nabla u|^2 - 2|K|^2 h_1^{-3} |\partial_2 \nabla u|^2 = 0 \quad \Rightarrow \quad h_1^2 = |K| \frac{|\partial_2 \nabla u|}{|\partial_1 \nabla u|},$$

and, consequently,

$$\frac{h_1}{h_2} \approx \frac{|\partial_2 \nabla u|}{|\partial_1 \nabla u|}.$$

In view of this result, we now consider the heuristic error indicator

$$\eta_K := \|\nabla(u - I_h u)\|_K \|\nabla(z - I_h z)\|_K,$$

which is minimized for

$$\frac{h_1}{h_2} \approx \frac{|\partial_2 \nabla u| \, |\partial_2 \nabla z|}{|\partial_1 \nabla u| \, |\partial_1 \nabla z|}. \tag{4.19}$$

This relation simultaneously reflects possible anisotropies in the primal and dual solution. The performance of anisotropic mesh adaptation based on the relations (4.17) and (4.19) will be compared in some numerical tests below.

*Remark* 4.14. We note that for the goal-oriented adaptation of purely tensor-product meshes an analogue of the mesh optimization strategy described in Section 4.2 can be developed, see Richter [123]. This approach yields meshes of optimal complexity in the case that the anisotropies of the primal and the dual solution are aligned with the coordinate directions. An extreme case occurs in Example 3.5, on a square domain, where the functional

$$J(u) = \int_{\partial\Omega} \partial_n u \, ds$$

results in a dual solution which is concentrated along the boundary $\partial\Omega$. Then, an optimal anisotropic refinement yields a mesh complexity of the order

$$TOL = \mathcal{O}(\sigma_h^{-1}),$$

for constant number $N$ of cells. This is much better than the complexity $TOL = \mathcal{O}(N^{-1})$, which can be achieved with isotropic meshes.

### Numerical test

As test case, we consider the Poisson problem on the square domain $\Omega = (-1, 1)^2$,

$$-\Delta u = f \quad \text{in } \Omega, \quad u_{|\partial\Omega} = 0, \tag{4.20}$$

for the exact solution

$$u(x) = (1 - x_1^2)^2 (1 - x_2)^2 (kx_1^2 + 0.1)^{-1},$$

where the parameter $k = 1, 4, 16, 64, \ldots$, determines the strength of the anisotropy. The right hand side is determined as $f := -\Delta u$. This solution is shown for $k = 4$ and $k = 64$ in Figure 4.7. The quantity to be computed is the mean value

$$J(u) := |\Omega|^{-1} \int_{\Omega} u \, dx.$$

In this case the anisotropy is only in the primal solution while the dual solution satisfies $-\Delta z = 1$ and is isotropic.

The computation starts from a coarse uniform tensor-product mesh which is then successively adapted on the basis of the relations (4.17) ('Strategy I') and (4.19) ('Strategy II'). The efficiency of the resulting meshes is depicted in Figure 4.8. Apparently, the meshes produced by Strategy II on the basis of the heuristic relation (4.19) are more efficient as the anisotropy increases than those obtained by Strategy I on the basis of the 'rigorous' relation (4.17).

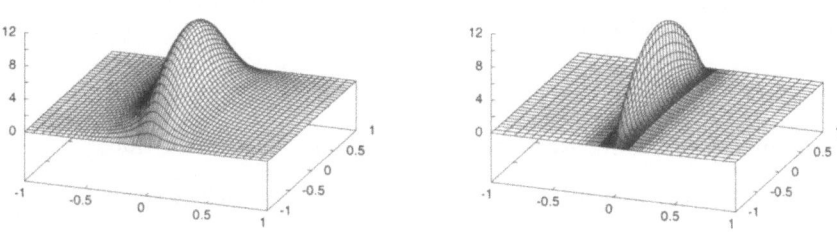

Figure 4.7: *Anisotropic solutions for $k = 4$ (left) and $k = 64$ (right); from Richter [123].*

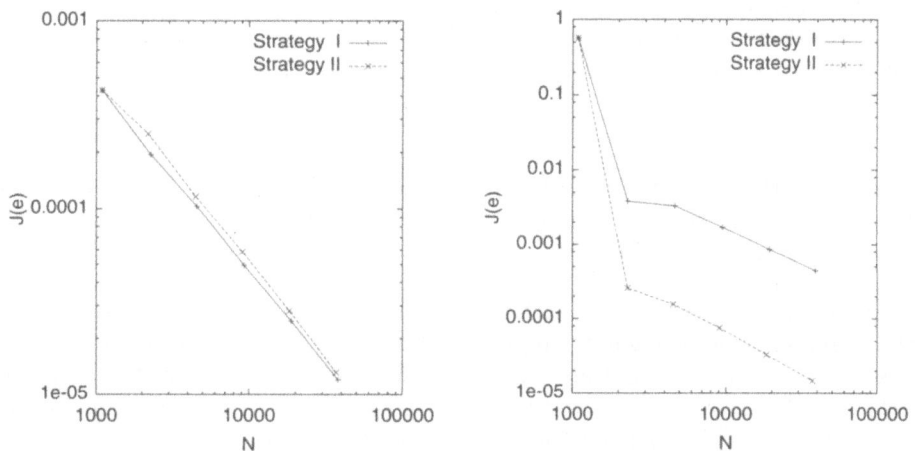

Figure 4.8: *Mesh efficiencies of anisotropic refinement by Strategy I and Strategy II, for $k = 1$ (left) and $k = 64$ (right); from Richter [123].*

## 4.5   Exercises

*Exercise* 4.1. The *fixed rate strategy* in a posteriori mesh adaptation refines and coarsens certain fractions $X$ and $Y$ of those cells with largest and smallest indicator values $\eta_K$, respectively,

$$\eta_{K_1} \geq \cdots \geq \eta_{K_i} \geq \cdots \geq \eta_{K_N}.$$

Design refining/coarsening strategies by specifying $X$, $Y$, such that

  a) the number $N$ of cells approximately doubles in each refinement cycle;

  b) the number $N$ of cells is approximately kept constant during the refinement process.

*Exercise* 4.2. The well-known Bramble-Hilbert theory guarantees the cell-wise error estimate

$$\|\nabla(u - I_h u)\|_K \leq c_I h_K \|\nabla^2 u\|_K$$

for the piecewise bilinear finite element nodal interpolation $I_h u$ on regular meshes, with a constant $c_I$ independent of $K$. Sketch in 2D the argument that this estimate has an analogue on cells with one or two *hanging nodes*,

$$\|\nabla(u - I_h u)\|_K \leq c_I h_K \|\nabla^2 u\|_{\tilde{K}},$$

where $\tilde{K}$ is the mother cell of $K$ on the preceding refinement level. Recall that at a hanging node the function value is set to be the average of the values at the two neighboring 'regular' nodes.

*Exercise* 4.3 *(Practical exercise).* Hanging nodes are commonly used to ease mesh refinement and coarsening in triangular or quadrilateral meshes. However, the presence of hanging nodes (and also transition cells) usually destroys the uniformity pattern of the mesh which may drastically reduce the accuracy of approximation. To demonstrate this, consider the model Poisson problem

$$-\Delta u = f \text{ in } \Omega, \quad u_{|\partial\Omega} = 0,$$

on $\Omega = (-1, 1)^2$, with right-hand side $f(x) = \frac{1}{2}\pi^2 \cos(\frac{1}{2}x_1\pi)\cos(\frac{1}{2}x_2\pi)$ and corresponding exact solution $u(x) = \cos(\frac{1}{2}x_1\pi)\cos(\frac{1}{2}x_2\pi)$. Compute the point-value $\partial_1 u(0.5, 0.5)$ on a sequence of

  a) uniform meshes with mesh-sizes $h = 2^{-1}$, $i = 1, 2, \ldots$;

  b) locally refined meshes using the a posteriori error estimator and the corresponding refinement strategy of Exercise 2.3.

Compare the achieved accuracy in terms of the number of cells $N$.

# Chapter 5

# The Limits of Theoretical Analysis

In this chapter, we want to discuss some questions concerning the theoretical justification of the DWR method for goal-oriented mesh adaptivity presented so far. We will see that this task is rather demanding and poses several new questions for the theoretical analysis of the finite element method. In fact, relying on the available results from the literature, we do not reach very far, yet. Since several not very practical assumptions will be used, we dispense with stating formal propositions.

To illustrate the problem, let us consider the special case of the evaluation of the derivative of a smooth solution $u \in C^2(\bar{\Omega})$ at some point $a \in \Omega \subset \mathbb{R}^2$. For this, we use the regularized output functional

$$J_\varepsilon(u) := |B_\varepsilon(a)|^{-1} \int_{B_\varepsilon(a)} \partial_1 u(a)\, dx = \partial_1 u(a) + \mathcal{O}(\varepsilon^2), \quad \varepsilon := TOL.$$

The corresponding dual solution behaves like a regularized 'derivative Green function' of the Laplacian:

$$|\nabla^k z(x)| \approx r(x)^{-k-1} := (|x-a|^2 + \varepsilon^2)^{-(k+1)/2}.$$

Then, for bilinear elements, the a posteriori error estimate takes the form

$$|J_\varepsilon(e)| \approx \eta_\omega := \sum_{K \in \mathbb{T}_h} \frac{h_K^3}{r_K^3}\, \rho_K, \qquad r_K := \max_{x \in K} r(x).$$

Now, assume that the local residuals are related to the local mesh size like

$$\rho_K \approx h_K, \tag{5.1}$$

uniformly for every $K \in \mathbb{T}_h$ and $h > 0$. This condition may be checked a posteriori in the course of the mesh adaptation process. Then, we obtain

$$\eta_\omega \approx \sum_{K \in \mathbb{T}_h} \frac{h_K^4}{r_K^3}. \tag{5.2}$$

In Section 4.2, we have seen that the optimal mesh for prescribed accuracy $TOL$ is characterized by the equilibration property

$$\eta_K = \frac{h_K^4}{r_K^3} \approx \frac{TOL}{N} \quad \Rightarrow \quad |J_\varepsilon(e)| \approx \sum_{K \in \mathbb{T}_h} \frac{TOL}{N} \approx \text{TOL}.$$

From this, we derive

$$h_K^2 \approx r_K^{3/2} \Big( \frac{TOL}{N} \Big)^{1/2},$$

and consequently,

$$N = \sum_{K \in \mathbb{T}_h} h_K^2 h_K^{-2} = \Big( \frac{N}{TOL} \Big)^{1/2} \sum_{K \in \mathbb{T}_h} h_K^2 r_K^{-3/2} \approx \Big( \frac{N}{TOL} \Big)^{1/2}.$$

This implies that

$$N \propto TOL^{-1}, \tag{5.3}$$

which is better than the $N \propto TOL^{-2}$ that could be achieved on uniformly refined meshes on the basis of the general *a priori* convergence estimate

$$|\partial_1 e(a)| = \mathcal{O}(h).$$

This predicted asymptotic behavior is well confirmed by the results shown in Table 5.1. We emphasize that in this example strong mesh refinement occurs, although the solution is smooth. In fact, this phenomenon should rather be interpreted as 'mesh coarsening' away from the point of evaluation. Further, observing that $r_{\min} \sim TOL$ and $r_{\max} \sim 1$, for the optimized mesh, there holds

$$h_{\min} \approx TOL^{5/4}, \quad h_{\max} \approx TOL^{1/2},$$

and, consequently, $h_{\min} \approx h_{\max}^{5/2}$, which means that

$$L \approx \frac{5}{4} \frac{\log(TOL)}{\log(2)}$$

refinement cycles are needed to reach the 'optimal' mesh.

Table 5.1: *Computing $\partial_1 u(0)$ using the estimator $\eta_\omega$ (L refinement levels); from Becker and Rannacher [30].*

| $TOL$ | $N$ | $L$ | $|J_\varepsilon(e)|$ | $\eta_\omega$ |
|-------|-----|-----|----------------------|---------------|
| $4^{-3}$ | 940 | 9 | $4.10 \cdot 10^{-1}$ | $1.42 \cdot 10^{-2}$ |
| $4^{-4}$ | 4912 | 12 | $4.14 \cdot 10^{-3}$ | $3.50 \cdot 10^{-3}$ |
| $4^{-5}$ | 20980 | 15 | $2.27 \cdot 10^{-4}$ | $9.25 \cdot 10^{-4}$ |
| $4^{-6}$ | 86740 | 17 | $5.82 \cdot 10^{-5}$ | $2.38 \cdot 10^{-4}$ |

*Remark* 5.1. Relation (5.1) is decisive for the optimality of a refined mesh. To see this, suppose that only

$$\rho_K \approx h_K^{1-\varepsilon}$$

holds, for some small $\varepsilon > 0$. Then, the above calculation would result in

$$h_K^2 \approx r_K^{6/(4-\varepsilon)} \left( \frac{TOL}{N} \right)^{2/(4-\varepsilon)},$$

and consequently,

$$N = \sum_{K \in \mathbb{T}_h} h_K^2 h_K^{-2} = \left( \frac{N}{TOL} \right)^{2/(4-\varepsilon)} \sum_{K \in \mathbb{T}_h} h_K^2 r_K^{-6/(4-\varepsilon)} \propto \left( \frac{N}{TOL} \right)^{2/(4-\varepsilon)}.$$

This would give us the asymptotic complexity

$$N \propto TOL^{-1-\varepsilon/2},$$

which growths faster than $TOL^{-1}$.

*Remark* 5.2. An alternative, more explicit strategy for mesh adaptation may be based on the balancing condition

$$\frac{h_K \rho_K}{r_K^3} \propto \frac{\text{TOL}}{|\Omega|} \quad \Rightarrow \quad |J_\varepsilon(e)| \propto \sum_{K \in \mathbb{T}_h} h_K^2 \frac{\text{TOL}}{|\Omega|} \approx \text{TOL}.$$

Here, the complexity analysis gives us (assuming again that $\rho_K \approx h_K$)

$$h_K \propto \left( \frac{\text{TOL}}{|\Omega|} \right)^{1/2} r_K^{3/2},$$

and, consequently,

$$N = \sum_{K \in \mathbb{T}_h} h_K^2 h_K^{-2} \propto \frac{|\Omega|}{\text{TOL}} \sum_{K \in \mathbb{T}_h} h_K^2 r_K^{-3} \propto \frac{|\Omega|}{\text{TOL}^2},$$

which shows that this approach is not efficient for very singular error functionals.

## 5.1   Convergence of residuals

The above example illustrates that the assumed asymptotic relation (5.1) for the residuals $\rho_K$ is crucial for deriving optimal complexity results. For analyzing this question in the case of $d$-linear finite elements, it suffices to consider the edge-residual part $\|[\partial_n u_h]\|_{\partial K}$, since on Cartesian meshes the cell-residual term automatically satisfies

$$\|f + \Delta u_h\|_K = \|f\|_K \leq c(f) h_K,$$

for bounded $f$. Now, notice that

$$h_K^{-3/2} \|[\partial_n u_h]\|_{\partial K} =: \left| D_h^2 u_{h|K} \right|$$

can be viewed as a mean value of a second-order difference quotient of $u_h$ on the cell-patch $\tilde{K}$ containing $K$ and its neighbors. Hence, in order to establish the bound (5.1), we have to seek for an estimate of the form

$$\sup_{h>0} \max_{x \in \Omega} |D_h^2 u_h| \leq c(u), \tag{5.4}$$

where the constant $c(u)$ depends on bounds for higher-order derivatives of the solution $u$. For proving (5.4), one may use the local inverse relation,

$$\|\nabla q\|_K \leq c_r h_K^{-1} \|q\|_K, \quad q \in P_r(K),$$

and the natural nodal interpolation $I_h u \in V_h$,

$$
\begin{aligned}
|D_h^2 u_{h|K}| &\leq h_K^{-1} \|D_h^2 u_h\|_K \\
&\leq h_K^{-1} \|D_h^2 (u_h - I_h u)\|_K + h_K^{-1} \|D_h^2 I_h u\|_K \\
&\leq c h_K^{-2} \|\nabla e\|_K + c h_K^{-2} \|\nabla(u - I_h u)\|_K + h_K^{-1} \|D_h^2 I_h u\|_K \\
&\leq c h_K^{-2} \|\nabla e\|_K + c h_K^{-1} \|\nabla^2 u\|_{\tilde{K}} \\
&\leq c \|\nabla^2 u\|_\infty,
\end{aligned}
$$

where again $\tilde{K}$ denotes a cell-patch neighborhood of $K$. Unfortunately, this argument only works on *quasi-uniform* meshes, for which $h_{\max}/h_{\min} \leq c$ is assumed, since the local error estimate

$$\|\nabla e\|_{\infty; K} \leq h_K c(u)$$

does not hold in this strong form on meshes with $h_{\max}/h_{\min} \to \infty$. What one can prove is the weaker version

$$\|\nabla e\|_{\infty; K} \leq c(u) \left\{ \max_{K' \in \hat{S}(K)} h_{K'} + h^2 \right\},$$

where $h := h_{max}$, and $\hat{S}(K)$ is some $\mathcal{O}(1)$-neighborhood of the cell $K$. Alternatives may be found by adopting weighted-norm techniques from the 'classical' pointwise error analysis of finite element approximation. However, most of these results also require the mesh to be quasi-uniform. Hence, the problem must be left open.

## 5.2 Approximation of weights

Next, we want to analyze the effect of approximating the dual solution $z$ in the weights $\omega_K$, on the accuracy of the error estimate. For simplicity, we again restrict our attention to the two-dimensional case. In the following, we will examine two different methods of approximating $z$:

*i) Approximation by a higher-order method.* First, we consider the approximation of the dual solution $z$ by its Ritz projection $z_h^{(2)}$ into the space $V_h^{(2)}$ of biquadratic finite elements, defined by

$$(\nabla \varphi_h, \nabla z_h^{(2)}) = J(\varphi_h) \quad \varphi_h \in V_h^{(2)}.$$

The resulting approximate error representation then reads

$$\tilde{E}(u_h) := \sum_{K \in \mathbb{T}_h} \left\{ (R_h, z_h^{(2)} - I_h z_h^{(2)})_K + (r_h, z_h^{(2)} - I_h z_h^{(2)})_{\partial K} \right\}.$$

Its difference to the exact error representation $E(u_h)$ can be written in the form

$$E(u_h) - \tilde{E}(u_h) = \rho(u_h)(\tilde{e}^* - I_h \tilde{e}^*) = (\nabla e, \nabla(\tilde{e}^* - I_h \tilde{e}^*)),$$

with the abbreviation $\tilde{e}^* := z - z_h^{(2)}$. For estimating this error, we recall the well-known a priori error estimate for biquadratic finite elements:

$$\left( \sum_{K \in \mathbb{T}_h} \|\nabla^k \tilde{e}^*\|_K^2 \right)^{1/2} \leq c h^{3-k} \|\nabla^3 z\|, \quad k = 0, 1, 2.$$

Using this, we can then estimate as follows:

$$
\begin{aligned}
|E(u_h) - \tilde{E}(u_h)| &\leq \|\nabla e\| \, \|\nabla(\tilde{e}^* - I_h \tilde{e}^*)\| \\
&\leq c h \|\nabla^2 u\| \left( \sum_{K \in \mathbb{T}_h} h_K^2 \|\nabla^2 \tilde{e}^*\|_K^2 \right)^{1/2} \\
&\leq c h^3 \|\nabla^2 u\| \, \|\nabla^3 z\|.
\end{aligned}
\tag{5.5}
$$

This estimate is unsatisfactory, as it requires the primal as well as the dual solution to be smooth which rules out most interesting applications. However, this objection can be somewhat weakened by the following modifications of the argument:

$$
\begin{aligned}
|E(u_h) - \tilde{E}(u_h)| &= |(\nabla e, \nabla \tilde{e}^*)| = |(\nabla(u - I_h u), \nabla \tilde{e}^*)| \\
&\leq c h^2 \left( \sum_{K \in \mathbb{T}_h} h_K^2 \|\nabla^2 u\|_K^2 \right)^{1/2} \|\nabla^3 z\|,
\end{aligned}
\tag{5.6}
$$

or, setting $\tilde{e} := u - u_h^{(2)}$ the 'biquadratic' Ritz projection error,

$$|E(u_h) - \tilde{E}(u_h)| = |(\nabla e, \nabla \tilde{e}^*)| = |(\nabla \tilde{e}, \nabla(z - I_h z))|$$

$$\leq ch^2 \|\nabla^3 u\| \Big( \sum_{K \in \mathbb{T}_h} h_K^2 \|\nabla^2 z\|_K^2 \Big)^{1/2}. \tag{5.7}$$

Here only smoothness of one of the solutions $u$ and $z$ is required and singularities in the other one can be compensated by proper mesh refinement. Though the estimates (5.6) and (5.7) are better than (5.5), they still do not cover point-error evaluation, i.e., $J(u) = u(x_0)$, since in this case

$$\sum_{K \in \mathbb{T}_h} h_K^2 \|\nabla^2 z\|_K^2 \approx \sum_{K \in \mathbb{T}_h} h_K^2 r_K^{-4} \approx \mathcal{O}(1),$$

and generally $h^2$ is not smaller than $TOL$. To get a better result, we may use a $L^\infty$-$L^1$-duality argument as follows:

$$|E(u_h) - \tilde{E}(u_h)| = |(\nabla e, \nabla \tilde{e}^*)| = |(\nabla(u - I_h^{(2)} u), \nabla \tilde{e}^*)|$$

$$\leq c \max_K \{h_K^2 \|\nabla^3 u\|_{\infty;K}\} \int_\Omega |\nabla \tilde{e}^*| \, dx \tag{5.8}$$

$$\leq ch|\log(h_{min})| \max_K \{h_K^2 \|\nabla^3 u\|_{\infty;K}\}.$$

The $L^1$-error estimate for the regularized Green function,

$$\int_\Omega |\nabla \tilde{e}^*| \, dx \leq c |\log(h_{min})|,$$

used in deriving (5.8) has been proven in Frehse and Rannacher [60]) only for quasi-uniform meshes; however, its extension to locally refined meshes is a technical exercise. The estimates (5.5)–(5.8) are useful provided that $h^3 \ll TOL$ on the current mesh. According to the discussion at the beginning, this is satisfied even in the case of derivative point value evaluation with

$$TOL \approx h_{max}^2.$$

However, since computing the dual solution by higher-order elements is too expensive in most practical situations, we will not pursue this discussion further.

*ii) Approximation by higher-order interpolation.* Next, we consider the theoretical justification of the approximate error estimator

$$\tilde{E}(u_h) := \sum_{K \in \mathbb{T}_h} \big\{ (R_h, I_{2h}^{(2)} z_h - z_h)_K + (r_h, I_{2h}^{(2)} z_h - z_h)_{\partial K} \big\},$$

where $I_{2h}^{(2)} z_h$ is the patch-wise *biquadratic* interpolation of the *bilinear* Ritz projection $z_h$ as defined above. This raises the question: *Why should $I_{2h}^{(2)} z_h$ be a better approximation to $z$ than $z_h$?* In fact, the construction of $I_{2h}^{(2)} z_h$ is based on nodal point information of $z_h$, and the point error $(z - z_h)(a)$ behaves generally not better then $\mathcal{O}(h^2)$, even on uniform meshes. Hence, it seems unlikely that

$$\|z - I_{2h}^{(2)} z_h\|_K \ll \|z - z_h\|_K.$$

However, this may not be the right point of view. We could also seek to prove the somewhat weaker relation

$$|\rho(u_h)(z - I_{2h}^{(2)} z_h)| \ll |\rho(u_h)(z - z_h)|,$$

which has the flavor of a global 'super-approximation' property. Therefore, it will probably depend on some uniformity property of the mesh. In order to pursue this thought further, we rewrite the error identity $J(e) = \rho(u_h)(z - z_h)$ in the form

$$J(e) = \rho(u_h)(z - I_{2h}^{(2)} z) + \rho(u_h)(I_{2h}^{(2)} z - I_{2h}^{(2)} z_h) + \rho(u_h)(I_{2h}^{(2)} z_h - z_h), \qquad (5.9)$$

where the last term on the right is just our approximate error estimator. The first and second term will be estimated separately. To this end, we have to assume that the meshes have been optimized such that

$$\rho_K \leq c h_K, \quad K \in \mathbb{T}_h. \qquad (5.10)$$

Using the local approximation properties of the interpolation $I_{2h}^{(2)} z$,

$$\left( \|z - I_{2h}^{(2)} z\|_K^2 + \tfrac{1}{2} h_K \|z - I_{2h}^{(2)} z\|_{\partial K}^2 \right)^{1/2} \leq c_I^{(2)} h_K^3 \|\nabla^3 z\|_{S(K)},$$

where $S(K)$ denotes the cell-patch on which $I_{2h}^{(2)} z$ is defined, we obtain for the first term in (5.9):

$$|\rho(u_h)(z - I_{2h}^{(2)} z)| \leq c \left( \sum_{K \in \mathbb{T}_h} h_K^6 \rho_K^2 \right)^{1/2} \|\nabla^3 z\|.$$

Consequently, using (5.10) we arrive at the estimate

$$|\rho(u_h)(z - I_{2h}^{(2)} z)| \leq c(u, z) h^3. \qquad (5.11)$$

The second term in (5.9) is the 'hard' one which requires more work, as it relates properties of the non-local Ritz projection and the local interpolation. Its estimation strongly relies on uniformity properties of the mesh $\mathbb{T}_h$. The idea is that the scaled error $h^{-2} e$ is a 'smooth' function, such that it can be approximated in $V_h$. To make this concept clear, suppose that the mesh $\mathbb{T}_h$ is *uniform* with mesh-width $h$. Then, it is known that in the nodal points, the error $z - z_h$ allows an asymptotic expansion in powers of $h$ which can be expressed in the form (see Blum and Rannacher [34])

$$I_h z - z_h = I_h(z - z_h) = h^2 I_h w + h^3 \tau_h,$$

with some $h$-independent function $w \in H_0^1(\Omega)$ and a remainder satisfying $\|\tau_h\| \le c\|\nabla^3 z\|$. From this, noting that $I_{2h}^{(2)} z = I_{2h}^{(2)} I_h z$, we conclude

$$\rho(u_h)(I_{2h}^{(2)} z - I_{2h}^{(2)} z_h) = h^2 \rho(u_h)(I_{2h}^{(2)} w) + h^3 \rho(u_h)(\tau_h),$$

and, using Galerkin orthogonality,

$$\rho(u_h)(I_{2h}^{(2)} z - I_{2h}^{(2)} z_h) = h^2 \rho(u_h)(I_{2h}^{(2)} w - I_h w) + h^3 \rho(u_h)(\tau_h).$$

Assuming that the interpolation operators $I_h$ and $I_{2h}^{(2)}$ behave like the $H^1$-stable Clément operator, we have

$$\|I_{2h}^{(2)} w - I_h w\|_K + \tfrac{1}{2} h^{1/2} \|I_{2h}^{(2)} w - I_h w\|_{\partial K} \le ch\|\nabla w\|_{S_K}.$$

Collecting the above estimates, and using again (5.10), we find

$$|\rho(u_h)(I_{2h}^{(2)} z - I_{2h}^{(2)} z_h)| \le c(u, z) h^3. \tag{5.12}$$

Finally, inserting the estimates (5.11) and (5.12) into (5.9), we conclude the desired estimate

$$J(e) = \tilde{E}(u_h) + \mathcal{O}(h^3). \tag{5.13}$$

We emphasize that this estimate has been derived for a smooth dual solution $z$ which excludes almost all interesting applications. Furthermore, the meshes are required to be uniform which conflicts with our ultimate goal of mesh adaptation. Therefore, a really meaningful analysis has to deal with the following three complications:

- The estimates are to be proven on locally refined meshes with only limited uniformity properties.

- The estimates are to be localized in order to allow for singularities in the dual solution.

- Cases of non-smooth solutions $u$, either due to data irregularities or due to reentrant corners (the more severe case) should be considered.

## 5.3 Exercises

*Exercise* 5.1. In the regularization of the output functional for the derivative point-value

$$J_\varepsilon(u) := |B_\varepsilon(a)|^{-1} \int_{B_\varepsilon(P)} \partial_1 u(a) \, dx = \partial_1 u(a) + \mathcal{O}(\varepsilon^2),$$

we have chosen $\varepsilon = \mathrm{TOL}$ which results in the mesh complexity $N \approx \mathrm{TOL}^{-1}$ and $h_{min} \approx h_{max}^{5/2} \approx \mathrm{TOL}^{5/4}$, i.e., the required number of refinement cycles becomes

$$L \approx \frac{5}{4} \frac{\log(\mathrm{TOL})}{\log(2)}.$$

For our goal $|\partial_1 e(a)| \approx \mathrm{TOL}$, it would suffice to choose $\varepsilon = \mathrm{TOL}^{1/2}$. What is the effect of this modification on the number $L$ of necessary refinement cycles.

*Exercise* 5.2. For the Galerkin finite element approximation of the Poisson problem by bilinear elements the a posteriori error estimate

$$|J(e)| \leq \eta_\omega := \sum_{K \in \mathbb{T}_h} \rho_K(u_h) \, \omega_K(z)$$

is derived from the error representation

$$\sum_{K \in \mathbb{T}_h} \left\{ (f + \Delta u_h, z - \psi_h)_K + \tfrac{1}{2}([\partial_n u_h], z - \psi_h)_{\partial K \setminus \partial \Omega} \right\}.$$

Show that by an appropriate choice of an interpolation $\psi_h = \tilde{I}_h z$ on patches of four cells, the cell-residual term can be made asymptotically smaller than the corresponding edge-residual term, provided that the right-hand side $f$ is sufficiently regular (say in $H^1(\Omega)$). Hence, in this case, the mesh refinement criterion can be based on the jump-residual terms alone. For a detailed analysis of this phenomenon see Carstensen and Verfürth [45].

*Exercise* 5.3 *(Practical exercise).* The critical assumption in estimating the efficiency of mesh refinement strategies based on weighted a posteriori error estimates is the 'convergence' of residuals, for bilinear finite elements expressed as

$$\|D_h^2 u_h\|_\infty = \max_{K \in \mathbb{T}_h} h_K^{-3/2} \|[\partial_n u_h]\|_K \leq c(\nabla^2 u),$$

if $u$ has uniformly bounded second derivatives, or more locally, for the case that $u$ is not regular (e.g., has corner singularities),

$$D_h^2 u_{h|K} := h_K^{-3/2} \|[\partial_n u_h]\|_K \leq c(\max_{\tilde{K}} |\nabla^2 u|),$$

with a cell patch $\tilde{K}$ around $K$, if $u$ is not regular (e.g., corner singularities) with constants independent of $h$. The first relation can be proven for quasi-uniform meshes, but is still open for general locally refined meshes including hanging nodes. We want to check these conditions by numerical experiments.

a) Consider the Poisson model problem

$$-\Delta u = f \text{ in } \Omega, \quad u_{|\partial\Omega} = 0,$$

on the domain $\Omega = (-1,1)^2$ with $f(x) = \frac{1}{2}\pi^2 \cos(\frac{1}{2}x_1\pi)\cos(\frac{1}{2}x_2\pi)$. Apply the duality-based mesh refinement strategy for computing the two different functional values:

$$J(u) = \partial_1 u(0.5, 0.5), \qquad J(u) = \int_{-1}^{1} \partial_1 u(1, x_2)\, dx_2,$$

and monitor the behavior of $D_h^2 u_{h|K}$ for increasingly refined meshes.

b) Consider the Poisson model problem

$$-\Delta u = f \text{ in } \Omega, \quad u_{|\partial\Omega} = 0,$$

on the slit-domain $\Omega = (-1,1)^2 \setminus \{x \in \mathbb{R}^2, x_1 = 0, -1 \le x_2 \le 0\}$ again with $f(x) = \frac{1}{2}\pi^2 \cos(\frac{1}{2}x_1\pi)\cos(\frac{1}{2}x_2\pi)$. Apply again the duality-based mesh refinement strategy to compute the same functional values as above, and monitor the behavior of the quantity

$$\|\tilde{D}_h^2 u_h\|_\infty := \max_{K\in\mathbb{T}_h} \left\{ r_K^{3/2} h_K^{-3/2} \|[\partial_n u_h]\|_K, \right\},$$

which reflects the expected singular behavior of $u$ at the tip of the slit, where $r_K$ is the distance of the cell $K$ from this point. Interpret the observed results.

# Chapter 6

# An Abstract Approach for Nonlinear Problems

In this chapter, we will present a very general approach to a posteriori error estimation for the Galerkin approximation of nonlinear variational problems as developed in Becker and Rannacher [31]. The framework is kept on an abstract level in order to allow later for a unified application to rather different situations, such as nonlinear PDEs, but also eigenvalue and optimization problems. We prepare for this by recalling the special situation of a linear Galerkin approximation of the form

$$a(u, \psi) = F(\psi) \qquad \forall \psi \in V, \tag{6.1}$$
$$a(u_h, \psi_h) = F(\psi_h) \quad \forall \psi_h \in V_h, \tag{6.2}$$

with a linear functional $F(\cdot)$ as right-hand side. For a given (linear) output functional $J(\cdot)$ the associated continuous and discrete dual problems are

$$a(\psi, z) = J(\psi) \qquad \forall \psi \in V, \tag{6.3}$$
$$a(\psi_h, z_h) = J(\psi_h) \quad \forall \psi_h \in V_h. \tag{6.4}$$

Then, using Galerkin orthogonality, for the 'primal' error $e := u - u_h$ and the 'dual' error $e^* := z - z_h$, there holds

$$J(e) = a(e, z) = a(e, e^*) = a(u, e^*) = F(e^*),$$

and, introducing the weighted *primal* and *dual* residuals,

$$J(e) = F(z - \psi_h) - a(u_h, z - \psi_h) =: \rho(u_h)(z - \psi_h), \qquad \psi_h \in V_h, \tag{6.5}$$
$$F(e^*) = J((u - \varphi_h) - a(u - \varphi_h, z_h) =: \rho^*(z_h)(u - \varphi_h), \quad \varphi_h \in V_h. \tag{6.6}$$

This trivially implies the 'linear' error representation

$$J(e) = \tfrac{1}{2}\rho(u_h)(z - \psi_h) + \tfrac{1}{2}\rho^*(z_h)(u - \varphi_h), \qquad \varphi_h, \psi_h \in V_h. \tag{6.7}$$

Below, we will see that this identity has a natural generalization to nonlinear problems, where the weighted primal and dual residuals are not identical.

## 6.1 Galerkin approximation of nonlinear equations

The following theory will be presented within an abstract functional analytic framework making as little assumptions as possible. Let $A(u)(\cdot)$ be a semilinear form and $J(\cdot)$ an output functional, not necessarily linear, defined on some function space $V$. The goal is the evaluation of $J(u)$ from the solution of the variational problem

$$A(u)(\psi) = 0 \quad \forall \psi \in V. \tag{6.8}$$

The corresponding Galerkin approximation uses finite dimensional subspaces $V_h \subset V$ to determine $u_h \in V_h$ by

$$A(u_h)(\psi_h) = 0 \quad \forall \psi_h \in V_h. \tag{6.9}$$

We assume the existence of directional derivatives of $A$ and $J$ up to order three denoted by $A'(u)(\varphi, \cdot)$, $A''(u)(\psi, \varphi, \cdot)$, $A'''(u)(\xi, \psi, \varphi, \cdot)$, and $J'(u)(\varphi)$, $J''(u)(\psi, \varphi)$, $J'''(u)(\xi, \psi, \varphi)$, respectively, for increments $\varphi, \psi, \xi \in V$. In these forms the dependence on the first argument in parentheses may be nonlinear while the dependence on all further arguments in the second set of parentheses is linear.

*Example* 6.1. A typical example of a nonlinear problem of the form we are interested in is the so-called *vector Burgers equation*

$$-\nu \Delta u + u \cdot \nabla u = f, \text{ in } \Omega, \quad u_{|\partial\Omega} = 0.$$

for a vector function $u \in H_0^1(\Omega)^d$. This is the natural generalization of the classical 1-D Burgers equation to multiple dimensions. The corresponding variational formulation has the form (6.8) with the semilinear form

$$A(u)(\psi) := \nu(\nabla u, \nabla \psi) + (u \cdot \nabla u, \psi) - (f, \psi).$$

*Example* 6.2. Another example of a nonlinear problem which will be considered in several exercises below is the diffusion-reaction equation

$$-\Delta u - u^3 = f, \text{ in } \Omega, \quad u_{|\partial\Omega} = 0.$$

for a scalar function $u \in H_0^1(\Omega)$. This problem is interesting in the context of bifurcation theory. In this case the corresponding semilinear form is given by

$$A(u)(\psi) := (\nabla u, \nabla \psi) - (u^3, \psi) - (f, \psi).$$

For estimating the error $J(u)-J(u_h)$, we employ the Euler-Lagrange method of constrained optimization. Introducing a 'dual' variable $z \in V$ ('adjoint' variable or 'Lagrangian multiplier'), we define the Lagrangian functional

$$\mathcal{L}(u,z) := J(u) - A(u)(z),$$

and seek for stationary points $\{u,z\} \in V \times V$ of $\mathcal{L}(\cdot,\cdot)$, i.e.,

$$\mathcal{L}'(u,z)(\varphi,\psi) = \left\{ \begin{matrix} J'(u)(\varphi) - A'(u)(\varphi,z) \\ -A(u)(\psi) \end{matrix} \right\} = 0 \quad \forall\{\varphi,\psi\}. \tag{6.10}$$

Clearly, the $u$-component of any such stationary point is a solution of the original problem (6.8). The Galerkin approximations $\{u_h, z_h\} \in V_h \times V_h$ are defined by the discrete Euler-Lagrange system

$$\mathcal{L}'(u_h,z_h)(\varphi_h,\psi_h) = \left\{ \begin{matrix} J'(u_h)(\varphi_h) - A'(u_h)(\varphi_h,z_h)) \\ -A(u_h)(\psi_h) \end{matrix} \right\} = 0 \quad \forall\{\varphi_h,\psi_h\}, \tag{6.11}$$

where, again, the $u_h$-component of any stationary point is a solution of the discrete problem (6.9). Now, the goal is to estimate the error $J(u)-J(u_h)$ in terms of the residuals associated with this set of equations. We prepare for this by considering first the general situation of the Galerkin approximation of stationary points of functionals.

**Proposition 6.1.** *Let $L(\cdot)$ be a three-times differentiable functional defined on a (real or complex) vector space $X$ which has a stationary point $x \in X$, i.e.,*

$$L'(x)(y) = 0 \quad \forall y \in X. \tag{6.12}$$

*Suppose that on a finite dimensional subspace $X_h \subset X$, the corresponding Galerkin approximation*

$$L'(x_h)(y_h) = 0 \quad \forall y_h \in X_h, \tag{6.13}$$

*has a solution $x_h \in X_h$. Then, there holds the error representation*

$$L(x) - L(x_h) = \tfrac{1}{2}L'(x_h)(x-y_h) + \mathcal{R}_h, \quad y_h \in X_h, \tag{6.14}$$

*with a remainder term $\mathcal{R}_h$ which is cubic in the error $e^x := x-x_h$,*

$$\mathcal{R}_h := \tfrac{1}{2} \int_0^1 L'''(x_h+se^x)(e^x,e^x,e^x)\, s(s-1)\, ds.$$

*Proof.* Since $L'(x)(e^x) = 0$, we can write

$$L(x) - L(x_h) = \int_0^1 L'(x_h+se^x)(e^x)\, ds$$
$$+ \tfrac{1}{2}L'(x_h)(e^x) - \tfrac{1}{2}\{L'(x_h)(e^x) + L'(x)(e^x)\}.$$

The last term on the right is just the approximation of the integral term by the trapezoidal rule. For this, we have the well-known error representation

$$\int_0^1 f(t)\,dt = \tfrac{1}{2}\{f(0) + f(1)\} + \tfrac{1}{2}\int_0^1 f''(s)s(s-1)\,ds.$$

Hence, we obtain

$$L(x) - L(x_h) = \tfrac{1}{2}L'(x_h)(e^x) + \tfrac{1}{2}\int_0^1 L'''(x_h+se^x)(e^x, e^x, e^x)\,s(s-1)\,ds.$$

Finally, observing that $L'(x_h)(y_h) = 0$ for all $y_h \in X_h$, we have that

$$L'(x_h)(e^x) = L'(x_h)(x - y_h), \quad y_h \in X_h.$$

This completes the proof.                                                                  □

As an immediate consequence of Proposition 6.1, we obtain the following result for the Galerkin approximation of variational equations.

**Proposition 6.2.** *For any solution of equations (6.8) and (6.9), we have the error representation*

$$J(u) - J(u_h) = \tfrac{1}{2}\rho(u_h)(z - \psi_h) + \tfrac{1}{2}\rho^*(u_h, z_h)(u - \varphi_h) + \mathcal{R}_h^{(3)}, \qquad (6.15)$$

*with arbitrary $\varphi_h, \psi_h \in V_h$, and the 'primal' and 'dual' residuals*

$$\rho(u_h)(\cdot) := -A(u_h)(\cdot),$$
$$\rho^*(u_h, z_h)(\cdot) := J'(u_h)(\cdot) - A'(u_h)(\cdot, z_h).$$

*The remainder term $\mathcal{R}_h^{(3)}$ is cubic in the 'primal' and 'dual' errors $e := u - u_h$ and $e^* := z - z_h$,*

$$\mathcal{R}_h^{(3)} = \tfrac{1}{2}\int_0^1 \{J'''(u_h+se)(e, e, e) - A'''(u_h+se)(e, e, e, z_h+se^*)$$
$$- 3A''(u_h+se)(e, e, e^*)\}\,s(s-1)\,ds.$$

*Proof.* In order to apply Proposition 6.1, we define the space $X := V \times V$ and for arguments $x = \{u, z\} \in X$ the functional $L(x) := \mathcal{L}(u, z)$. In this context the stationary points are denoted by $x := \{u, z\}$ and $x_h := \{u_h, z_h\}$ with the error $e^x := x - x_h$. Then, we have

$$J(u) - J(u_h) = L(x) - A(u)(z) - L(x_h) + A_h(u_h)(z_h) = L(x) - L(x_h).$$

Hence, the error representation of Proposition 6.1 gives us

$$J(u) - J(u_h) = \tfrac{1}{2}L'(x_h)(x - y_h) + \mathcal{R}_h, \quad y_h \in X_h,$$

with

$$\mathcal{R}_h := \tfrac{1}{2} \int_0^1 L'''(x_h + se^x)(e^x, e^x, e^x)\, s(s-1)\, ds.$$

By construction, we have for arbitrary $y_h = \{\varphi_h, \psi_h\} \in X_h$ :

$$
\begin{aligned}
L'(x_h)(x - y_h) &= \mathcal{L}'_u(u_h, z_h)(u - \varphi_h) + \mathcal{L}'_z(u_h, z_h)(z - \psi_h) \\
&= J'(u_h)(u - \varphi_h) - A'(u_h)(u - \varphi_h, z_h) - A(u_h)(z - \psi_h) \\
&= \rho^*(u_h, z_h)(u - \varphi_h) + \rho(u_h)(z - \psi_h).
\end{aligned}
$$

Notice that $\mathcal{L}(u, z)$ is linear in $z$. Consequently, the third derivative of $L(\cdot)$ consists of only three terms, namely,

$$J'''(u_h + se)(e, e, e) - A'''(u_h + se)(e, e, e, z_h + se^*) - 3A''(u_h + se)(e, e, e^*).$$

This implies the asserted form of the remainder term $\mathcal{R}_h^{(3)}$. □

*Remark 6.3.* The derivation of the error representations (6.14) and (6.15) does not require the uniqueness of solutions; this is important, for example, for the application to eigenvalue problems. In cases with non-unique solutions, the a priori assumption $x_h \to x$ ($h \to 0$) makes the result meaningful as then the remainder term can be assumed to be small.

*Remark 6.4.* The actual evaluation of the error identity requires guesses for the primal and dual solutions $u$ and $z$ which are usually computed from the Galerkin approximations $u_h$ and $z_h$ by some post-processing as described in Section 4.1.

*Remark 6.5.* The cubic remainder $\mathcal{R}_h^{(3)}$ can usually be neglected. However, in parameter-dependent problems when approaching a bifurcation point, the derivatives of $A(u)(\cdot)$ and consequently $\mathcal{R}_h^{(3)}$ may become large. In such a situation the abstract theory can still be applied but with special care. The extreme situation will be seen in Chapter 7 in the context of eigenvalue problems where one is directly working in the singular case.

In the linear case, we have seen that the primal and dual residual terms coincide, i.e. $\rho(u_h)(z - \psi_h) = \rho^*(z_h)(u - \varphi_h)$. This is no longer true in the nonlinear case, but the deviation from this property, i.e. the degree of nonlinearity of the problem, can be estimated as the following proposition shows.

**Proposition 6.6.** *With the notation from above, there holds*

$$\rho^*(u_h, z_h)(u - \varphi_h) = \rho(u_h)(z - \psi_h) + \Delta\rho, \qquad (6.16)$$

*for any* $\varphi_h, \psi_h \in V_h$ *, with*

$$\Delta\rho := \int_0^1 \left\{ A''(u_h + se)(e, e, z_h + se^*) - J''(u_h + se)(e, e) \right\} ds.$$

*Further, we have the simplified error representation*

$$J(u) - J(u_h) = \rho(u_h)(z-\varphi_h) + \mathcal{R}_h^{(2)}, \tag{6.17}$$

*for any $\varphi_h \in V_h$, with the quadratic remainder*

$$\mathcal{R}_h^{(2)} := \int_0^1 \{A''(u_h+se)(e,e,z) - J''(u_h+se)(e,e)\}\, s\, ds$$

*Proof.* We introduce the scalar function $g(\cdot)$ by

$$g(s) := J'(u_h+se)(e) - A'(u_h+se)(e, z_h+se^*).$$

By the definition of $z$ and $z_h$, there holds

$$g(1) = J'(u)(e) - A'(u)(e, z) = 0,$$
$$g(0) = J'(u_h)(e) - A'(u_h)(e, z_h) = \rho^*(u_h, z_h)(e),$$

and

$$g'(s) = J''(u_h+se)(e, e) - A''(u_h+se)(e, e, z_h+se^*)$$
$$\quad - A'(u_h+se)(e, e^*).$$

Therefore, using Galerkin orthogonality,

$$\rho^*(u_h, z_h)(u-\varphi_h) = \rho^*(u_h, z_h)(e) = g(0) = g(0) - g(1) = -\int_0^1 g'(s)\, ds$$

$$= \int_0^1 \{A''(u_h+se)(e, e, z_h+se^*) - J''(u_h+se)(e, e)\}\, ds$$

$$+ \int_0^1 A'(u_h+se)(e, e^*)\, ds$$

$$= \Delta\rho + \rho(u_h)(e^*) = \Delta\rho + \rho(u_h)(z-\psi_h).$$

this proves (6.16). In order to prove (6.17), we use integration by parts, obtaining

$$\mathcal{R}_h^{(2)} = \int_0^1 \{A''(u_h+se)(e, e, z) - J''(u_h+se)(e, e)\}\, s\, ds$$

$$= -\int_0^1 \{A'(u_h+se)(e, z) - J'(u_h+se)(e)\}\, ds + A'(u)(e, z) - J'(u)(e),$$

where the last two terms vanish by definition of $z$. Consequently, employing again Galerkin orthogonality, we obtain

$$\mathcal{R}_h^{(2)} = -\rho(u_h)(z-\psi_h) + J(u) - J(u_h),$$

for arbitrary $\psi_h \in V_h$. This completes the proof.
    We note that the simplified error representation (6.17) could have been derived also from (6.15) using the relation (6.16). However, this involves lengthy calculations, so that we preferred to present a more direct argument.   □

In order to use the error representations (6.15) or (6.17) for practical mesh adaptation, we have to evaluate the primal and dual residual terms. As in the linear case, this requires approximation of the dual solution $z$ and in the context of (6.15) additionally that of the primal solution $u$. This may be achieved again by post-processing of the Galerkin solutions $z_h$ and $u_h$ exploiting higher-order interpolation. Let the resulting approximations be denoted by $\tilde{z}_h$ and $\tilde{u}_h$. Then, neglecting the remainder terms the approximate error representations take the form

$$\tilde{E}(u_h, z_h) := \tfrac{1}{2}\rho(u_h)(\tilde{z}_h - z_h) + \tfrac{1}{2}\rho^*(u_h, z_h)(\tilde{u}_h - u_h), \qquad (6.18)$$

and

$$\tilde{E}(u_h) := \rho(u_h)(\tilde{z}_h - z_h). \qquad (6.19)$$

*Remark* 6.7. The identity (6.16) is useful as it offers the possibility of controlling the remainder $\mathcal{R}_h^{(2)}$ in the simplified error representation (6.17). In fact, comparing the two error representations (6.15), (6.17) and using (6.16), we see that

$$
\begin{aligned}
\mathcal{R}_h^{(2)} &= -\rho(u_h)(z - \psi_h) + J(u) - J(u_h) \\
&= -\rho(u_h)(z - \psi_h) + \tfrac{1}{2}\rho(u_h)(z - \psi_h) + \tfrac{1}{2}\rho^*(u_h, z_h)(u - \varphi_h) + \mathcal{R}_h^{(3)} \\
&= \tfrac{1}{2}\rho^*(u_h, z_h)(u - \varphi_h) - \tfrac{1}{2}\rho(u_h)(z - \psi_h) + \mathcal{R}_h^{(3)} \\
&= \tfrac{1}{2}\Delta\rho + \mathcal{R}_h^{(3)}.
\end{aligned}
$$

Hence, we may try to control the linearization error by a posteriori checking the condition

$$|\Delta\rho| \approx |\rho^*(u_h, z_h)(\tilde{u}_h - u_h) - \rho(u_h)(\tilde{z}_h - z_h)| \ll TOL. \qquad (6.20)$$

where $\tilde{u}_h \approx u$ and $\tilde{z}_h \approx z$ are higher-order approximations and the cubic remainder term $\mathcal{R}_h^{(3)}$ is neglected.

*Remark* 6.8. The possibility of improving on the *linearization error*, i.e. reducing the remainder term $\mathcal{R}_h^{(2)}$, by post-processing the Ritz approximation $u_h$ has been studied in Vexler [134]. This costly process may be relevant in cases when the problem to be solved is close to bifurcation.

*Remark* 6.9. A posteriori error estimates for the Galerkin finite element approximation of nonlinear variational problems have also been derived using extensions of the classical 'energy-norm-based' approach; see, e.g., Verfürth [131, 133]. These results rely on assumptions on the monotonicity of the underlying problem, i.e. the coercivity of certain derivative forms, or involve nonlinear stability constants which depend on the unknown solution. These estimates usually represent the 'worst case' scenario and will only in special cases be of practical value.

## 6.2   A nested solution approach

For solving the nonlinear problems by a Galerkin finite element method, we employ the following iterative scheme. Starting from a coarse initial mesh $\mathbb{T}_0$, a hierarchy of refined meshes

$$\mathbb{T}_0 \subset \mathbb{T}_1 \subset \cdots \subset \mathbb{T}_l \subset \cdots \subset \mathbb{T}_L,$$

and corresponding finite element spaces $V_l$, $l = 0, ..., L$, with dimensions $N_l$ is generated by the following nested solution process:

1. *Initialization:* For $l = 0$, compute a solution $u_0 \in V_0$ on the mesh $\mathbb{T}_0$.

2. *Defect correction iteration:* For $l \geq 1$, start with $u_l^{(0)} = u_{l-1} \in V_l$. For a computed iterate $u_l^{(j)} \in V_l$ evaluate the defect

$$(d_l^{(j)}, \psi_l) = -A(u_l^{(j)})(\psi_l), \quad \forall \psi_l \in V_l,$$

and solve the correction equation (Newton update)

$$\tilde{A}'(u_l^{(j)})(v_l^{(j)}, \psi_l) = (d_l^{(j)}, \psi_l) \quad \forall \psi_l \in V_l,$$

by Krylov-space or multigrid iterations using the hierarchy of previously constructed meshes $\{\mathbb{T}_{l-1}, ..., \mathbb{T}_0\}$. Update $u_l^{(j+1)} = u_l^{(j)} + \alpha_l^{(j)} v_l^{(j)}$, with a step-length parameter $\alpha_l^{(j)}$, set $j = j+1$, and repeat the iteration. This process is carried until a limit $u_l \in V_l$, is reached with some required accuracy.

3. *Error estimation:* Solve the (linearized) discrete dual problem

$$z_l \in V_l : \quad A'(u_l)(\varphi_l, z_l) = J(\varphi_l) \quad \forall \varphi_l \in V_l,$$

and evaluate the a posteriori error estimate (6.18):

$$J(e_l) \approx \tilde{E}(u_l, z_l).$$

If $|\tilde{E}(u_l, z_l)| \leq TOL$, or $N_l \geq N_{\max}$, then stop. Otherwise cell-wise mesh adaptation yields the new mesh $\mathbb{T}_{l+1}$. Then, set $l := l + 1$ and go to (2).

*Remark 6.10.* The nonlinear iteration described above is oriented at the solution of stationary elliptic problems in which a global transfer of information is present. In the case of transport-dominated problems, particularly those with information transfer into one direction only such as in nonstationary problems, one would organize the solution process differently, taking this transport direction into account.

*Remark 6.11.* In the described Newton-like iteration the mesh adaptation is done for the limit solution on the current mesh in order to have a rigorous theoretical basis. However, it may be inefficient to carry the iteration on a coarser mesh to the limit knowing that the discretization accuracy on this mesh is still insufficient. Hence, one would like to combine the estimation of the discretization error with that of the iteration error, both in accordance with the accuracy in the target quantity. Such a combined error estimator has been developed for a linear multigrid iteration in Becker et al. [27] and Becker [19], but it is an open questions for the nonlinear Newton iteration.

*Remark* 6.12. The solution of the *linear* dual problem usually requires much less work compared to solving the nonlinear primal problem (6.8). In fact, in the context of the nested solution method described above the primal solution is obtained by a Newton iteration which requires the solution of several linear problems until convergence is reached. Then, solving the linear dual problem for the converged primal solution normally corresponds to about one additional Newton step. Further, this extra work is spent on optimized meshes adapted to the particular goal of the computation. Hence, particularly for nonlinear problems, the duality-based approach to adaptivity becomes relatively 'cheap'. This is demonstrated by the examples from structural and fluid mechanics which will be presented in Chapter 10 and Chapter 11.

*Remark* 6.13. We remark on the following particular aspect in the evaluation of the dual residual $\rho^*(u_h, z_h)(u-\varphi_h)$. The discrete dual solution $z_h$ is determined by the variational equation (6.11) which is the Galerkin discretization of the corresponding continuous variational problem (6.10). The latter can usually be interpreted as the 'weak' form of a certain system of differential equations. The cell and edge residuals $R_h^*$ and $r_h^*$ of the dual solution $z_h$ are then taken with respect to the 'strong' form of this dual problem. However, in deriving these residuals, we have to observe their consistency with the variational formulation (6.11) in order not to miss terms which may not be present on the continuous level, such as for example the residual terms $\nabla \cdot v_{h|K}$ occurring in solving the incompressible Navier-Stokes equations. We will comment on this point later on when a 'naive' derivation of the dual cell and edge residuals may lead to wrong results (see Remark 11.2).

## 6.3 Exercises

*Exercise* 6.1. Consider the Galerkin approximation of the stationary 1-dimensional Burgers equation

$$-\nu u_{xx} + u u_x = f, \quad \text{in } (0,1), \quad u(0) = 0 = u(1),$$

by using piecewise linear finite elements. Determine the cubic and quadratic remainder terms in the a posteriori error representations for a *linear* output functional.

*Exercise* 6.2. Use the second-order 'nonlinear' a posteriori error estimate to derive the $L^2$-norm a posteriori error bound for the Galerkin finite element approximation of the (linear) Poisson problem.

*Exercise* 6.3 *(Practical exercise).* Consider the nonlinear diffusion-reaction equation

$$-\Delta u - u^3 = f, \quad \text{in } \Omega, \quad u_{|\partial\Omega} = 0,$$

where $\Omega$ is the domain defined in Exercise 3.3. The discretization is by the usual bilinear finite elements. For $f \equiv \alpha = 1, \dots, 75$, compute the solution's mean value over the subdomain $\Omega' = \{x \in \Omega, \, x_1 > .5, x_2 > .5\} \subset \Omega$,

$$J_1(u) := 4 \int_{\Omega'} u(x)\, dx.$$

Monitor the size (relative to the estimated error) of the control term for the linearization error,

$$\Delta\tilde{\rho} := \rho^*(u_h, z_h)(\tilde{I}_{2h}^{(2)} u_h - u_h) - \rho(u_h)(\tilde{I}_{2h}^{(2)} z_h - z_h),$$

by approximating the weights using the *biquadratic patch-wise interpolations* $\tilde{I}_{2h}^{(2)} u_h$ and $\tilde{I}_{2h}^{(2)} z_h$ of the discrete primal and dual solutions $u_h$ and $z_h$, respectively. Repeat the same calculations using the true *biquadratic Ritz projection* $z_h^{(2)}$ of $z$ instead. Explain the observed results.

# Chapter 7

# Eigenvalue Problems

In the following, we will apply the abstract theory of the DWR method developed in Chapter 6 to error control in the approximation of eigenvalue problems. We mention some prototypical examples we are particularly interested in:

- The *symmetric* eigenvalue problem of the Laplace operator:

$$-\Delta u = \lambda u.$$

- The *nonsymmetric* eigenvalue problem of a convection-diffusion operator:

$$-\Delta u + b \cdot \nabla u = \lambda u.$$

- The *stability* eigenvalue problem governed by the linearized Navier-Stokes equations:

$$-\nu \Delta v + \hat{v} \cdot \nabla v + v \cdot \nabla \hat{v} + \nabla p = \lambda v, \quad \nabla \cdot v = 0,$$

  where $\hat{v}$ is some 'base solution' the stability of which is to be investigated.

The results we will present for the approximation of these problems are taken from Heuveline and Rannacher [77, 78].

The above eigenvalue problems are all treated within an abstract setting, which will be laid out in the following. Let $V$ be a (complex) Hilbert space. We seek $\{u, \lambda\} \in V \times \mathbb{C}$ satisfying

$$\mathcal{A}u = \lambda \mathcal{M}u,$$

with linear operators $\mathcal{A}$ and $\mathcal{M}$ in $V$. The operator $\mathcal{M}$ is assumed to be self-adjoint and positive semi-definite. It is introduced to allow for the situation of the stability eigenvalue problem for the Navier-Stokes equations, where the eigenvalue term only occurs in the first of the two equations (the momentum equation). We do not go deeper into the abstract functional analytic setting of eigenvalue problems since we will concentrate below on concrete problems formulated in the standard function spaces. Again, we prefer the variational formulation

$$a(u, \psi) = \lambda \, m(u, \psi) \quad \forall \psi \in V, \tag{7.1}$$

where $a(\cdot, \cdot) := \langle \mathcal{A} \cdot, \cdot \rangle$ is the (generally nonsymmetric) sesquilinear form generated by $\mathcal{A}$ and $m(\cdot, \cdot)$ is the symmetric semi-definite sesquilinear form corresponding to $\mathcal{M}$ (usually the $L^2$ scalar product). Eigenfunctions are assumed to be normalized by $m(u, u) = 1$. For nonsymmetric $\mathcal{A}$, one also considers the associated *adjoint eigenvalue problem* for the Hilbert-space adjoint $\mathcal{A}^*$ of $\mathcal{A}$,

$$\mathcal{A}^* u^* = \lambda^* \mathcal{M} u^*.$$

Observing that $a^*(u^*, \varphi) = \langle \mathcal{A}^* u^*, \varphi \rangle = \overline{a(\varphi, u^*)}$ and $m(u^*, \varphi) = \overline{m(\varphi, u^*)}$, this has the variational form

$$a(\varphi, u^*) = \bar{\lambda}^* \, m(\varphi, u^*) \quad \forall \varphi \in V. \tag{7.2}$$

Note that in the context of matrix eigenvalue problems the *dual eigenvectors* $u^*$ are also called *left eigenvectors*. From the definition, we see that primal and dual eigenvalues are related by $\lambda^* = \bar{\lambda}$, while the corresponding eigenvectors may differ. Usually, the dual eigenvectors are normalized by requiring

$$m(u, u^*) = 1.$$

This normalization is possible only if $u^*$ is not $m$-orthogonal with respect to the whole eigenspace of $\lambda$, which is equivalent to requiring that $\lambda$ has trivial *defect*, i.e., its algebraic eigenspace reduces to the geometric one (only trivial Jordan blocks in the matrix case); for a more detailed discussion of this issue see Heuveline and Rannacher [77], and the literature cited therein. We will see that the simultaneous consideration of primal and dual eigenvalue problems is essential for rigorous a posteriori error estimation.

The Galerkin approximation of (7.1) and (7.2) uses finite dimensional subspaces $V_h \subset V$, as described above, to determine $\{u_h, \lambda_h\}, \{u_h^*, \lambda_h^*\} \in V_h \times \mathbb{C}$, satisfying

$$a(u_h, \psi_h) = \lambda_h \, m(u_h, \psi_h) \quad \forall \psi_h \in V_h, \quad m(u_h, u_h) = 1, \tag{7.3}$$

$$a(\varphi_h, u_h^*) = \bar{\lambda}_h^* \, m(\varphi_h, u_h^*) \quad \forall \varphi_h \in V_h, \quad m(u_h, u_h^*) = 1. \tag{7.4}$$

Our goal is to control the errors $\lambda - \lambda_h$, $u - u_h$, and $u^* - u_h^*$ in the eigenvalues and eigenfunctions in terms of the residuals associated with these equations.

## 7.1   A posteriori error analysis

In order to derive a posteriori error estimates, we embed the present situation into the general framework of variational equations as considered in Chapter 6.

Consider the product spaces $\mathcal{V} := V \times \mathbb{C}$ and $\mathcal{V}_h := V_h \times \mathbb{C} \subset \mathcal{V}$ and define for pairs $U := \{u, \lambda\} \in \mathcal{V}$ and $\Psi = \{\psi, \nu\} \in \mathcal{V}$ the semilinear form

$$A(U)(\Psi) := \lambda\, m(u, \psi) - a(u, \psi) + \bar{\nu}\, \{m(u, u) - 1\}.$$

With this notation, the above continuous eigenvalue problem and its discrete analogue can be written in the following compact form of semilinear variational equations:

$$A(U)(\Psi) = 0 \qquad \forall \Psi \in \mathcal{V}, \tag{7.5}$$
$$A(U_h)(\Psi_h) = 0 \qquad \forall \Psi_h \in \mathcal{V}_h, \tag{7.6}$$

where $U := \{u, \lambda\}$ and $U_h := \{u_h, \lambda_h\}$, respectively. In order to control the error in the approximation of the eigenvalues, we use the output functional

$$J(\Phi) := \mu\, m(\varphi, \varphi).$$

Since $m(u, u) = 1$ at the solution $U = \{u, \lambda\}$, there holds

$$J(U) = \lambda\, m(u, u) = \lambda,$$

i.e., as desired this functional picks out the eigenvalue. We recall the Lagrange approach from Chapter 6, particularly the Lagrangian functional for arguments $U = \{u, \lambda\}$ and $\Psi = \{\psi, \nu\}$:

$$\begin{aligned}
\mathcal{L}(U, \Psi) &= J(U) - A(U)(\Psi) \\
&= \lambda\, m(u, u) - \lambda\, m(u, \psi) + a(u, \psi) - \bar{\nu}\, \{m(u, u) - 1\}.
\end{aligned}$$

The dual solution $Z = \{z, \pi\} \in \mathcal{V}$ and its Galerkin approximation $Z_h = \{z_h, \pi_h\} \in \mathcal{V}_h$ are then determined by the equations

$$A'(U)(\Phi, Z) = J'(U)(\Phi) \qquad \forall \Phi \in \mathcal{V}, \tag{7.7}$$
$$A'(U_h)(\Phi_h, Z_h) = J'(U_h)(\Phi_h) \qquad \forall \Phi_h \in \mathcal{V}_h. \tag{7.8}$$

The left and right hand sides of (7.7) read, for $Z = \{z, \pi\}$, $U = \{u, \lambda\}$, and $\Phi = \{\varphi, \mu\}$, as follows:

$$\begin{aligned}
A'(U)(\Phi, Z) &= \lambda\, m(\varphi, z) - a(\varphi, z) + \mu\, m(u, z) + 2\bar{\pi}\,\mathrm{Re}\, m(\varphi, u), \\
J'(U)(\Phi) &= \mu\, m(u, u) + 2\lambda\,\mathrm{Re}\, m(\varphi, u).
\end{aligned}$$

Hence, the continuous dual problem takes the form

$$\lambda\, m(\varphi, z) - a(\varphi, z) + \mu\{m(u, z) - m(u, u)\} + 2\{\bar{\pi} - \lambda\}\mathrm{Re}\, m(\varphi, u) = 0,$$

for all $\Phi = \{\varphi, \mu\}$. We now show that the dual eigenpair $U^* = \{u^*, \lambda^*\}$ solves this equation, i.e., $Z = \{z, \pi\} = U^*$ is one solution. This is clear since for $m(u, z) = m(u, u) = 1$ and $\bar{\pi} = \lambda$, the equation reduces to

$$a(\varphi, z) = \bar{\pi}\, m(\varphi, z),$$

which is the defining equation for $U^*$. Whether this is the only solution is not relevant for the present discussion. Analogously, the discrete adjoint problem (7.8) is solved by the discrete dual eigenpair $U_h^* = \{u_h^*, \lambda_h^*\}$ determined by

$$a(\varphi_h, u_h^*) = \bar{\lambda}_h^* \, m(\varphi_h, u_h^*) \quad \forall \varphi_h \in V_h, \quad m(u_h, u_h^*) = 1. \tag{7.9}$$

With these preparations, we can formulate the following proposition:

**Proposition 7.1.** *With the primal and dual eigenvalue residuals*

$$\rho(u_h, \lambda_h)(\cdot) := a(u_h, \cdot) - \lambda_h \, m(u_h, \cdot),$$
$$\rho^*(u_h^*, \lambda_h^*)(\cdot) := a(\cdot, u_h^*) - \bar{\lambda}_h^* \, m(\cdot, u_h^*),$$

*we have the error representation*

$$\lambda - \lambda_h = \tfrac{1}{2}\rho(u_h, \lambda_h)(u^* - \psi_h) + \tfrac{1}{2}\rho^*(u_h^*, \lambda_h^*)(u - \varphi_h) + \mathcal{R}_h, \tag{7.10}$$

*for arbitrary* $\psi_h, \varphi_h \in V_h$, *with the cubic remainder term*

$$\mathcal{R}_h = \tfrac{1}{2}(\lambda - \lambda_h) \, m(u - u_h, u^* - u_h^*).$$

*Proof.* The assertion is an immediate consequence of Proposition 6.2 applied to the present situation. We have

$$J(U) - J(U_h) = \tfrac{1}{2}\big\{ J'(U_h)(U - \Phi_h) - A'(U_h)(U - \Phi_h, Z_h) \big\}$$
$$+ \tfrac{1}{2}\big\{ - A(U_h)(Z - \Psi_h) \big\} + \mathcal{R}_h,$$

for arbitrary $\Phi_h = \{\varphi_h, \mu\}$, $\Psi_h = \{\psi_h, \chi\} \in \mathcal{V}_h$, with the cubic remainder term $\mathcal{R}_h$ which we evaluate below. Hence, using the above preparations,

$$\lambda - \lambda_h = \tfrac{1}{2}\big\{ (\lambda - \mu) \, m(u_h, u_h) + 2\lambda_h \, \mathrm{Re} \, m(u - \varphi_h, u_h)$$
$$+ a(u - \varphi_h, z_h) - \lambda_h \, m(u - \varphi_h, z_h) - (\lambda - \mu) \, m(u_h, z_h)$$
$$- 2\bar{\pi}_h \, \mathrm{Re} \, m(u - \varphi_h, u_h) - (\lambda - \mu)\{m(u_h, u_h) - 1\} \big\}$$
$$- \tfrac{1}{2}\big\{ \lambda_h \, m(u_h, z - \psi_h) - a(u_h, z - \psi_h) + (\bar{\pi} - \bar{\chi})\{m(u_h, u_h) - 1\} \big\}$$
$$+ \mathcal{R}_h.$$

We are free to choose $\mu := \lambda$, $\chi := \pi$, and observing that $m(u, u) = m(u_h, u_h) = 1$, and $\lambda_h = \bar{\pi}_h = \bar{\lambda}_h^*$, we obtain

$$\lambda - \lambda_h = \tfrac{1}{2}\big\{ a(u - \varphi_h, z_h) - \lambda_h \, m(u - \varphi_h, z_h) \big\}$$
$$+ \tfrac{1}{2}\big\{ a(u_h, z - \psi_h) - \bar{\lambda}_h^* \, m(u_h, z - \psi_h) \big\} + \mathcal{R}_h.$$

It remains to evaluate the remainder term. Setting $E := \{u - u_h, \lambda - \lambda_h\}$ and $E^* := \{u^* - u_h^*, \lambda^* - \lambda_h^*\}$, the general remainder term from Proposition 6.2 is

$$\mathcal{R}_h = \tfrac{1}{2} \int_0^1 \big\{ J'''(U_h + sE)(E, E, E) - A'''(U_h + sE)(E, E, E, Z_h + sE^*)$$
$$- 3A''(U_h + sE)(E, E, E^*) \big\} s(s - 1) \, ds.$$

In the present case, by a simple calculation, we have

$$J'''(U_h+sE)(E,E,E) = 6(\lambda-\lambda_h)m(u-u_h,u-u_h),$$
$$A'''(U_h+sE)(E,E,E,Z_h+sE^*) = 0,$$
$$-3A''(U_h+sE)(E,E,E^*) = -6(\lambda-\lambda_h)\,m(u-u_h,u^*-u_h^*)$$
$$- 6(\bar{\lambda}^*-\bar{\lambda}_h^*)\,m(u-u_h,u-u_h).$$

Consequently, noting that $\lambda-\lambda_h = \bar{\lambda}^*-\bar{\lambda}_h^*$,

$$\mathcal{R}_h = -3\int_0^1 (\lambda-\lambda_h)\,m(u-u_h,u^*-u_h^*)s(s-1)\,ds$$
$$= \tfrac{1}{2}(\lambda-\lambda_h)\,m(u-u_h,u^*-u_h^*),$$

which completes the proof. $\qquad\qquad\qquad\qquad\qquad\qquad\qquad\qquad\qquad\square$

*Remark* 7.2. We add the following remarks concerning Proposition 7.1:

- The error representation (7.10) holds true without any assumption on the multiplicity of the eigenvalue $\lambda$ or its defect. However, such a restriction will become necessary below in dealing with the error of the eigenfunctions.

- In the error representation (7.10) only terms involving the computed primal and dual eigenpairs occur and no additional outer dual problem needs to be solved. We will see a similar situation in the context of optimization problems discussed in Chapter 8 where the underlying mechanism will become clear.

- In the nonsymmetric case the simultaneous consideration of primal and dual eigenvalue problems is essential within an optimal multigrid iteration anyway (see Heuveline and Bertsch [76]). Then, the computation of $u^*$ for the error estimator does therefore not introduce extra work.

- In Proposition 7.1, we have assumed that the governing operator $\mathcal{A}$ remains unchanged under discretization, i.e. all coefficient are frozen. Below, in considering the stability eigenvalue problem, we will additionally allow for approximation of the operator $\mathcal{A}(\hat{u})$ depending on coefficients $\hat{u}$ that will also be subject to approximation.

## Practical evaluation of the error representation

Next, we want to determine the explicit form of the residuals in Proposition 7.1. To this end, we need to be more specific about the particular structure of the eigenvalue problem considered. Here, we restrict ourselves to a simple model situation which, nevertheless, is prototypical for the problems we are interested in. On a polygonal or polyhedral domain $\Omega \subset \mathbb{R}^d$ consider the eigenvalue problem of a second-order elliptic differential operator $\mathcal{A}$ such as, for example,

$$\mathcal{A}v := -\Delta v + b{\cdot}\nabla v = \lambda v \quad \text{in } \Omega, \quad \mathrm{v}_{|\partial\Omega} = 0,$$

with a smooth (or even constant) transport coefficient $b$. In this case, we have $\mathcal{M} := \mathrm{id}$. Further, let this eigenvalue problem be approximated by the Galerkin method using piecewise linear or $d$-linear finite elements on meshes $\mathbb{T}_h = \{K\}$, as described in Chapter 3. Within this setting, we can proceed analogously as before, obtaining

$$
\begin{aligned}
\rho(u_h, \lambda_h)(\cdot) &= a(u_h, \cdot) - \lambda_h\, m(u_h, \cdot) \\
&= \sum_{K \in \mathbb{T}_h} \left\{ (\mathcal{A}u_h - \lambda_h \mathcal{M}u_h, \cdot)_K - (\partial_n^{\mathcal{A}} u_h, \cdot)_{\partial K} \right\} \\
&= \sum_{K \in \mathbb{T}_h} \left\{ (\mathcal{A}u_h - \lambda_h \mathcal{M}u_h, \cdot)_K + \tfrac{1}{2}([\partial_n^{\mathcal{A}} u_h], \cdot)_{\partial K} \right\},
\end{aligned}
$$

$$
\begin{aligned}
\rho^*(u_h^*, \lambda_h^*)(\cdot) &= a(\cdot, z_h) - \lambda_h^*\, m(\cdot, z_h) \\
&= \sum_{K \in \mathbb{T}_h} \left\{ (\cdot, \mathcal{A}^* z_h - \lambda_h^* \mathcal{M} z_h)_K - (\cdot, \partial_n^{\mathcal{A}^*} z_h)_{\partial K} \right\} \\
&= \sum_{K \in \mathbb{T}_h} \left\{ (\cdot, \mathcal{A}^* z_h - \lambda_h^* \mathcal{M} z_h)_K + \tfrac{1}{2}(\cdot, [\partial_n^{\mathcal{A}^*} z_h])_{\partial K} \right\}.
\end{aligned}
$$

Hence, using again the notation of 'equation' and 'jump residuals' $R_h$, $R_h^*$, $r_h$, and $r_h^*$, respectively, analogously as introduced in Section 3.1, the residual admits the estimate

$$
|\rho(u_h, \lambda_h)(u^* - \psi_h) + \rho^*(u_h^*, \lambda_h^*)(u - \varphi_h)| \leq \sum_{K \in \mathbb{T}_h} \left\{ \rho_K \omega_K^* + \rho_K^* \omega_K \right\}, \qquad (7.11)
$$

with the cell residuals $\rho_K$, $\rho_K^*$ and and weights $\omega_K$, $\omega_K^*$ defined by

$$
\begin{aligned}
\rho_K &:= \left( \|R_h\|_K^2 + h_K^{-1/2} \|r_h\|_{\partial K}^2 \right)^{1/2}, \\
\rho_K^* &:= \left( \|R_h^*\|_K^2 + h_K^{-1/2} \|r_h^*\|_{\partial K}^2 \right)^{1/2}, \\
\omega_K &:= \left( \|u - \varphi_h\|_K^2 + \tfrac{1}{2} h_K^{1/2} \|u - \varphi_h\|_{\partial K}^2 \right)^{1/2}, \\
\omega_K^* &:= \left( \|u^* - \psi_h\|_K^2 + h_K^{1/2} \|u^* - \psi_h\|_{\partial K}^2 \right)^{1/2}.
\end{aligned}
$$

As a consequence of the above discussion, we obtain the following result:

**Proposition 7.3.** *Within the above setting, assuming that*

$$
|m(u - u_h, u^* - u_h^*)| \leq 1, \qquad (7.12)
$$

*we have the 'weighted' a posteriori error estimate*

$$
|\lambda - \lambda_h| \leq \eta_\lambda^\omega := \sum_{K \in \mathbb{T}_h} \left\{ \rho_K \omega_K^* + \rho_K^* \omega_K \right\}, \qquad (7.13)
$$

*and the 'energy-norm-error-type' estimate*

$$|\lambda - \lambda_h| \le \eta_\lambda^{(1)} := c_\lambda \sum_{K \in \mathbb{T}_h} h_K^2 \{\rho_K^2 + \rho_K^{*2}\}, \tag{7.14}$$

*with a constant $c_\lambda$ growing linearly with $|\lambda|$.*

*Proof.* Using the estimate (7.11) in the error representation (7.10) gives us

$$|\lambda - \lambda_h| \le \tfrac{1}{2} \sum_{K \in \mathbb{T}_h} \{\rho_K \omega_K^* + \rho_K^* \omega_K\} + |\mathcal{R}_h|.$$

Since, in virtue of assumption (7.12),

$$|\mathcal{R}_h| = \tfrac{1}{2}|(\lambda - \lambda_h)\, m(u - u_h, u^* - u_h^*)| \le \tfrac{1}{2}\,|\lambda - \lambda_h|,$$

the asserted estimate (7.13) follows. To prove (7.14), we choose $\psi_h := \tilde{I}_h u^*$ and $\varphi_h := \tilde{I}_h u$ in $\omega_K^*$ and $\omega_K$, with the modified nodal interpolation operator $\tilde{I}_h$ introduced in Section 3.2. Writing

$$u^* - \tilde{I}_h u^* = (u^* - u_h^*) - \tilde{I}_h(u^* - u_h^*), \quad u - \tilde{I}_h u = (u - u_h) - \tilde{I}_h(u - u_h),$$

in the weights $\omega_K^*$ and $\omega_K$, we obtain by the interpolation estimate (3.10) that

$$\sum_{K \in \mathbb{T}_h} h_K^{-2} \{\omega_K^{*2} + \omega_K^2\} \le \tilde{c}_I^2 \{\|\nabla(u^* - u_h^*)\|^2 + \|\nabla(u - u_h)\|^2\}.$$

This gives us

$$|\lambda - \lambda_h| \le \tfrac{1}{2}\tilde{c}_I \Big( \sum_{K \in \mathbb{T}_h} h_K^2 \{\rho_K^2 + \rho_K^{*2}\} \Big)^{1/2} \big(\|\nabla(u^* - u_h^*)\|^2 + \|\nabla(u - u_h)\|^2\big)^{1/2}.$$

Now, the proof would be completed by showing that the energy-norm errors of the eigenfunctions are proportionally bounded by the eigenvalue error. This is actually the case but requires lengthy calculations employing duality arguments for the eigenfunction errors. These details are omitted and we refer instead to Heuveline and Rannacher [77]. $\qquad\square$

*Remark 7.4.* The analogue of the a posteriori error estimator $\eta_\lambda^{(1)}$ for *symmetric* eigenvalue problems has been given by Nystedt [108]. There, the symmetry of the problem is extensively used. Furthermore, $H^2$-regularity of the eigenfunctions is required which excludes domains with reentrant corners, the most interesting case for an adaptive approach. Notice that our derivation of $\eta_\lambda^{(1)}$ does not need these assumptions.

*Remark* 7.5. An alternative version of the eigenvalue-error estimator $\eta_\lambda^{(1)}$ has been given by Larson [101], also for the symmetric case and assuming $H^2$-regularity of the eigenfunctions, namely,

$$|\lambda - \lambda_h| \leq \eta_\lambda^{(2)} := c_\lambda \Big( \sum_{K \in \mathbb{T}_h} h_K^4 \{\rho_K^2 + \rho_K^{*2}\} \Big)^{1/2}. \qquad (7.15)$$

Both error estimators, $\eta_\lambda^{(1)}$ and $\eta_\lambda^{(2)}$, are asymptotically equivalent for regular situations. However, the required $H^2$ regularity for the eigenfunctions renders the estimator $\eta_\lambda^{(2)}$ useless for typical situations in which mesh adaptivity is needed.

*Remark* 7.6. Neglecting the presence of the *dual* eigenvalue problem, we may try to control the error in approximating the eigenvalue using the *primal* residual part alone, i.e. using the 'reduced' error estimator

$$\eta_\lambda^{\mathrm{red}} := \sum_{K \in \mathbb{T}_h} h_K^2 \rho_K^2.$$

Below, we will compare the performance of the above eigenvalue-error estimators $\eta_\lambda^\omega$, $\eta_\lambda^{(1)}$, $\eta_\lambda^{(2)}$, and $\eta_\lambda^{\mathrm{red}}$ for a simple model situation.

## Numerical examples

We consider the convection-diffusion model eigenvalue problem

$$-\Delta v + b\cdot\nabla v \;=\; \lambda v \quad \text{in } \Omega, \quad v_{|\partial\Omega} = 0,$$

on the slit-domain $\Omega = (-1,1)\times(-1,3) \setminus \{x \in \mathbb{R}^2, x_1 = 0, -1 < x_2 \leq 0\}$, with the transport vector $b = (0,b_y)^T$; for a sketch of this configuration, see Figure 7.1. In the computations on this test problem the mesh refinement is organized according to the 'fixed-rate' strategy with refinement rate $X = 0.2$, see Section 4.2.

### Test 1: Symmetric case

At first, we consider the symmetric eigenvalue problem, i.e. $b = 0$. The eigenfunction corresponding to the smallest eigenvalue and the computational mesh generated on the basis of the weighted error estimator $\eta_\lambda^\omega$ are shown in Figure 7.1. Figure 7.2 shows the mesh efficiencies achieved on the basis of the different error estimators $\eta_\lambda^\omega$, $\eta_\lambda^{(1)}$ and $\eta_\lambda^{(2)}$ introduced above compared with that of uniform mesh refinement. (In this case $\eta_\lambda^{\mathrm{red}}$ and $\eta_\lambda^{(1)}$ are equivalent.) We see that in the symmetric case all estimators show almost equally good performance compared to that of uniform refinement. This is due to the dominance of the error caused by the slit singularity which is well captured by all residual-based error estimators. In fact, it is known that in the symmetric case, the eigenvalue error is proportional to the square of the energy-norm error.

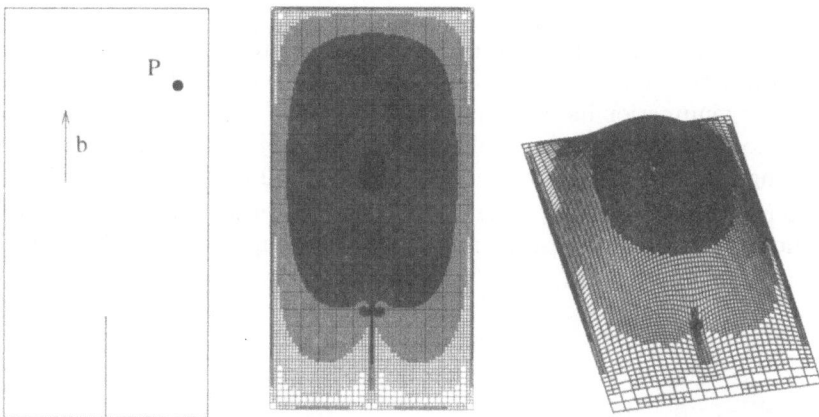

Figure 7.1: *Configuration for* $b \equiv 0$ *(left), adapted mesh with* 12,000 *cells (middle), eigenfunction (right); from Heuveline and Rannacher [77].*

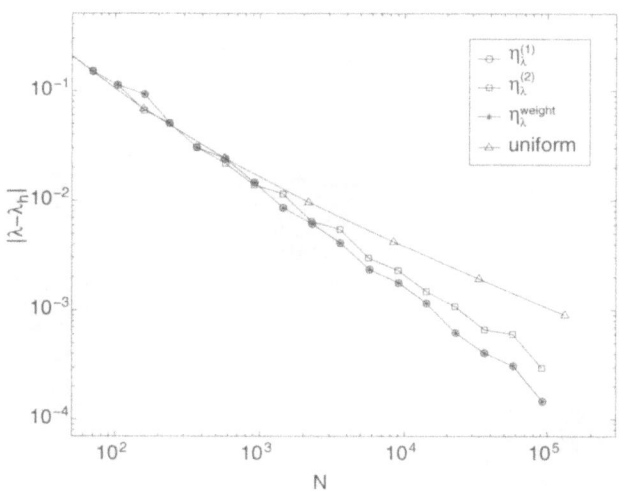

Figure 7.2: *Mesh efficiency achieved using the different error estimators,* $\eta_\lambda^{(1)}$ *(symbol ○),* $\eta_\lambda^{(2)}$ *(symbol □), and* $\eta_\lambda^\omega$ *(symbol ∗), compared against uniform refinement (symbol △). The curves for* $\eta_\lambda^{(1)}$ *and* $\eta_\lambda^\omega$ *lie above each other; from Heuveline and Rannacher [77].*

**Test 2: Nonsymmetric case**

Next, we consider the nonsymmetric version of the test eigenvalue problem with vertical transport, $b_y = 3$. In this case, due to the Dirichlet boundary conditions, the primal eigenfunction has a steeper gradient at the top boundary, while the dual eigenfunction has one at the bottom boundary; see Figure 7.3. The latter boundary layer strongly interferes with the slit singularity. Figure 7.4 shows adapted meshes obtained using the 'energy-type' error estimator $\eta_\lambda^{(1)}$, the 'reduced' error estimator $\eta_\lambda^{red}$, and the 'weighted' error estimator $\eta_\lambda^\omega$. The superiority of the latter one is clearly seen in Figure 7.5.

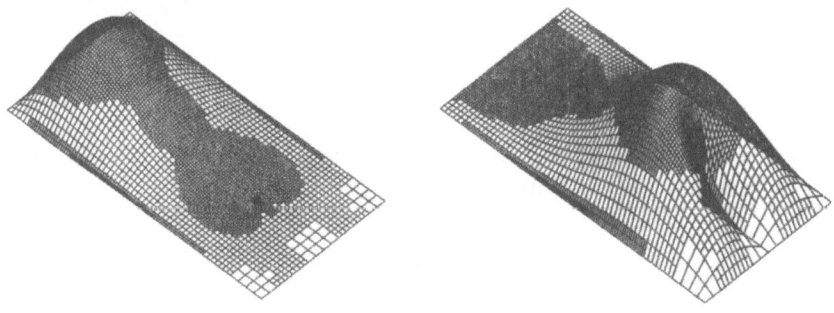

Figure 7.3: *Primal (left) and dual eigenfunction (right); from Heuveline and Rannacher [77].*

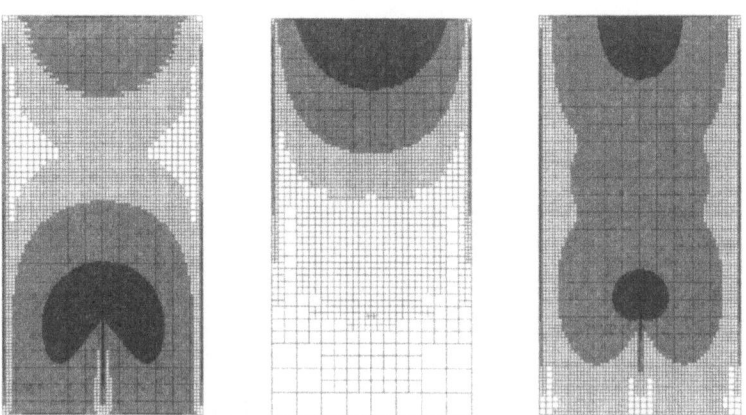

Figure 7.4: *Adapted meshes with* $10,000$ *cells on the basis of the error estimators* $\eta_\lambda^{(1)}$ *(left),* $\eta_\lambda^{red}$ *(middle),* $\eta_\lambda^\omega$ *(right); from Heuveline and Rannacher [77].*

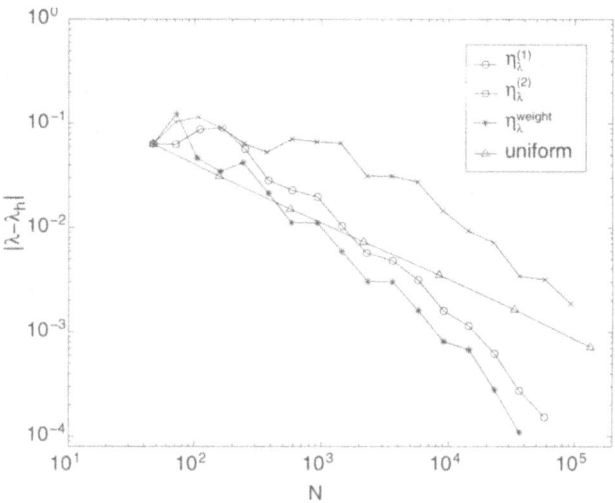

Figure 7.5: *Mesh efficiencies achieved on the basis of the different error estimators: $\eta_\lambda^{(1)}$ (symbol $\circ$), $\eta_\lambda^{red}$ (symbol $\times$), and $\eta_\lambda^\omega$ (symbol $*$), compared against uniform refinement (symbol $\triangle$); from Heuveline and Rannacher [77].*

## 7.2 Error control for functionals of eigenfunctions

We now turn to the question of how to estimate the error with respect to output functionals of the eigenfunctions. This question seems non-trivial since at an eigenvalue the corresponding adjoint operator is naturally singular, so that it is not clear what the proper definition of the dual problem has to be in this case.

Let $j(\cdot) : V \to \mathbb{C}$ be an output functional, for simplicity assumed to be linear, with respect to which the error in the eigenfunctions is to be controlled. For ease of presentation, we also assume that the eigenvalue $\lambda$ considered is simple with a normalized eigenfunction $u$. In order to apply our abstract theory, we define the functional

$$J(\Phi) := j(\varphi), \quad \Phi = \{\varphi, \mu\} \in \mathcal{V}.$$

Then, the associated abstract dual problem for $Z = \{z, \pi\} \in \mathcal{V}$,

$$A'(U)(\Phi, Z) = J'(U)(\Phi) \qquad \forall \Phi \in \mathcal{V},$$

takes the explicit form

$$\lambda\, m(\varphi, z) - a(\varphi, z) + \mu\, m(u, z) - 2\,\bar{\pi}\, \mathrm{Re}\, m(\varphi, u) = j(\varphi), \qquad (7.16)$$

for all  $\Phi = \{\varphi, \mu\} \in \mathcal{V}$ . This is equivalent to

$$\lambda m(\varphi, z) - a(\varphi, z) = j(\varphi) + 2\,\bar{\pi}\,\mathrm{Re}\,m(\varphi, u) \quad \forall \varphi \in V,$$

and  $m(u, z) = 0$ . Since  $\lambda$  is assumed to be a simple eigenvalue, by virtue of the Fredholm alternative, this equation can be solved if and only if its right-hand side vanishes on the eigenvector  $u$ , that is

$$j(u) - 2\,\bar{\pi}\,\mathrm{Re}\,m(u, u) = 0 \quad \Leftrightarrow \quad \bar{\pi} = \tfrac{1}{2}\,j(u).$$

Consequently, for the given eigenpair  $\{u, \lambda\}$  the so reduced dual problem

$$a(\varphi, z) - \lambda\,m(\varphi, z) = j(\varphi) - j(u)\,\mathrm{Re}\,m(\varphi, u) \quad \forall \varphi \in V, \qquad (7.17)$$

has a solution  $z \in V$ , which is uniquely determined in view of the property  $m(u, z) = 0$ . By an analogous argument, the reduced discrete dual problem is seen to be

$$a(\varphi_h, z_h) - \lambda_h\,m(\varphi_h, z_h) = j(\varphi_h) - j(u_h)\,\mathrm{Re}\,m(\varphi_h, u_h) \quad \forall \varphi_h \in V_h, \qquad (7.18)$$

where  $\{u_h, \lambda_h\}$  is the eigenpair of the approximate eigenvalue problem, and  $\bar{\pi}_h := \tfrac{1}{2}\,j(u_h)$ . This problem also has a solution  $z_h \in V_h$ , uniquely determined by  $m(u_h, z_h) = 0$ . We see that the well-posedness of these dual problems is guaranteed by filtering out the eigenspaces  $\mathrm{span}\{u\}$  and  $\mathrm{span}\{u_h\}$ , respectively. With all this, we can state the following proposition.

**Proposition 7.7.** *Let  $\{u_h, \lambda_h\}$  be a computed eigenpair approximating  $\{u, \lambda\}$ . Then, for the given functional  $j(\cdot) : V \to \mathbb{C}$  and the associated solution  $z \in V$  of the dual problem (7.17), we have the error representation*

$$\begin{aligned} j(u - u_h) &= \rho(u_h, \lambda_h)(z - \psi_h) + (\lambda - \lambda_h)\,m(u - u_h, z) \\ &\quad + \tfrac{1}{2}j(u)\,m(u - u_h, u - u_h), \end{aligned} \qquad (7.19)$$

*for arbitrary  $\psi_h \in V_h$ .*

*Proof.* First, we recall the definitions of the primal (eigenvalue) residual

$$\rho(u_h, \lambda_h)(\cdot) = a(u_h, \cdot) - \lambda_h m(u_h, \cdot),$$

and the dual residual associated to the dual problem (7.17),

$$\rho^*(u_h, z_h)(\cdot) := a(\cdot, z_h) - \lambda_h\,m(\cdot, z_h) + j(\cdot) - j(u_h)\,\mathrm{Re}\,m(\cdot, u_h).$$

This implies with the results of Chapter 6 that

$$\begin{aligned} j(u - u_h) &= \tfrac{1}{2}\left\{ a(u_h, z - \psi_h) - \lambda_h m(u_h, z - \psi_h) \right\} \\ &\quad + \tfrac{1}{2}\left\{ a(u - \varphi_h, z_h) - \lambda_h m(u - \varphi_h, z_h) \right. \\ &\quad \left. - j(u_h)\,\mathrm{Re}\,m(u - \varphi_h, u_h) + j(u - \varphi_h) \right\} + \mathcal{R}_h, \end{aligned}$$

and consequently, taking $\varphi_h = u_h$,

$$j(u-u_h) = a(u_h, z-\psi_h) - \lambda_h m(u_h, z-\psi_h) + a(u-u_h, z_h)$$
$$- \lambda_h m(u-u_h, z_h) - j(u_h)\,\text{Re}\,m(u-u_h, u_h) + 2\mathcal{R}_h.$$

To identify the remainder $\mathcal{R}_h$, we note that

$$J'''(U_h + sE; E, E, E) = 0,$$
$$A'''(U_h + sE; Z_h + sE^*)(E, E, E) = 0,$$
$$-3A''(U_h + sE; E, E, E^*) = -6(\lambda - \lambda_h)m(u-u_h, z-z_h)$$
$$- 6(\bar{\pi} - \bar{\pi}_h)m(u-u_h, u-u_h),$$

which yields

$$\mathcal{R}_h = \tfrac{1}{2}(\lambda - \lambda_h)m(u-u_h, z-z_h) + \tfrac{1}{2}(\bar{\pi} - \bar{\pi}_h)m(u-u_h, u-u_h).$$

We recall that $\bar{\pi} = \tfrac{1}{2}j(u)$ and $\bar{\pi}_h = \tfrac{1}{2}j(u_h)$ and obtain

$$\mathcal{R}_h = \tfrac{1}{2}(\lambda - \lambda_h)m(u-u_h, z-z_h) + \tfrac{1}{4}j(u-u_h)m(u-u_h, u-u_h).$$

From this, we infer as an intermediate result:

$$j(u-u_h) = a(u_h, z-\psi_h) - \lambda_h m(u_h, z-\psi_h)$$
$$+ a(u-u_h, z_h) - \lambda_h m(u-u_h, z_h)$$
$$- j(u_h)\,\text{Re}\,m(u-u_h, u_h) + \tfrac{1}{2}j(u-u_h)m(u-u_h, u-u_h)$$
$$+ (\lambda - \lambda_h)m(u-u_h, z-z_h).$$

Next, by definition and since $m(u_h, z_h) = 0$, we have

$$a(u-u_h, z_h) - \lambda_h m(u-u_h, z_h) = (\lambda - \lambda_h)m(u, z_h) = (\lambda - \lambda_h)m(u-u_h, z_h).$$

Further, noting that $m(u, u) = m(u_h, u_h) = 1$, there holds

$$m(u-u_h, u-u_h) = m(u, u) + m(u_h, u_h) - 2\,\text{Re}\,m(u, u_h)$$
$$= m(u, u) - m(u_h, u_h) - 2\,\text{Re}\,m(u - u_h, u_h)$$
$$= -2\,\text{Re}\,m(u - u_h, u_h).$$

Then, combining the last three relations gives us

$$j(u-u_h) = a(u_h, z-\psi_h) - \lambda_h m(u_h, z-\psi_h) + (\lambda - \lambda_h)m(u-u_h, z_h)$$
$$+ \tfrac{1}{2}j(u)\,m(u-u_h, u-u_h) + (\lambda - \lambda_h)m(u-u_h, z-z_h),$$

which completes the proof. $\qquad\square$

*Remark* 7.8. The proposition requires $\lambda$ to be simple. In the case of geometric multiplicity $\rho > 1$, we have to simultaneously consider a whole basis $\{u^{(i)}, i = 1, ..., \rho\}$ of the eigenspace $\text{kern}(\mathcal{A} - \lambda I)$ in setting up the dual problem. The case of higher algebraic multiplicity can also be handled but is much more involved.

## Numerical example

In order to illustrate the foregoing result, we consider the model problem from above. The goal is to evaluate the derivative value $j(u) := \partial_1 u(a)$ of the 'first' eigenfunction at $a = (0.5, 2.5)^T$ .

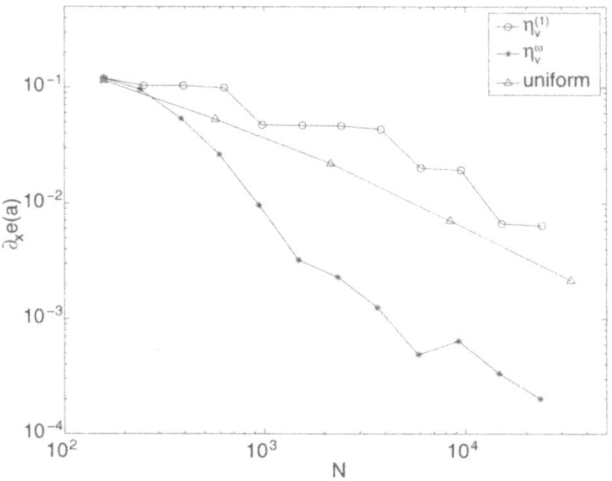

Figure 7.6: *Mesh efficiency of $\eta_u^\omega$ (symbol $*$) compared to $\eta_u^{(1)}$ (symbol O) and uniform refinement (symbol $\Delta$); from Heuveline and Rannacher [77].*

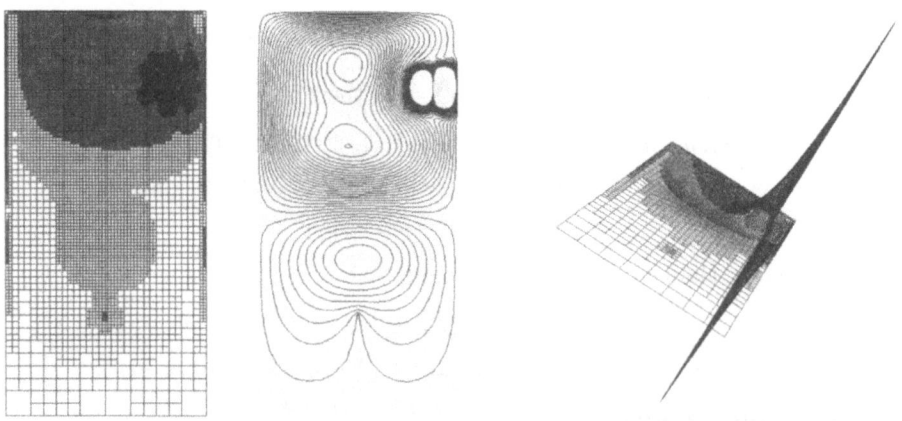

Figure 7.7: *Adapted mesh with about $12,000$ cells for the approximation of $\partial_1 u(a)$ obtained by $\eta_u^\omega$ (left), and the corresponding dual solution (middle and right); from Heuveline and Rannacher [77].*

## 7.3 The stability eigenvalue problem

Eigenvalue problems play an important role in the analysis of the *stability* of solutions of nonlinear differential equations. Classical areas of applications are structural mechanics and fluid mechanics, in the latter context referred to as *hydrodynamic stability*. We will consider this kind of problem again within an abstract setting. Let $a(\cdot)(\cdot)$ be a given semilinear form determining the base solution $\hat{u} \in V$ by

$$a(\hat{u})(\psi) = 0 \quad \forall \psi \in V, \tag{7.20}$$

and its Galerkin approximation $\hat{u}_h \in V_h \subset V$ by

$$a(\hat{u}_h)(\psi_h) = 0 \quad \forall \psi_h \in V_h. \tag{7.21}$$

Note that $\hat{u}$ is a *stationary* solution of the system under consideration. We want to determine whether this base solution is *dynamically stable*, i.e., whether any non-stationary solution trajectory $\{\tilde{u}(t) \in V, t \geq 0\}$, starting from a small perturbation $\tilde{u}(0) = \hat{u} + w^0$, and satisfying the evolution equation associated to (7.20),

$$m(\partial_t \tilde{u}, \psi) + a(\tilde{u})(\psi) = 0 \quad \forall \psi \in V, \tag{7.22}$$

stays bounded or even decays back to $\hat{u}$, as $t \to \infty$. Here, the decay is expressed in the semi-norm $m(\cdot, \cdot)^{1/2}$, which in most practical cases is actually the usual $L^2$ norm. However, in order to prepare for the application of this approach also to the incompressible Navier-Stokes equations in Chapter 11, we allow $m(\cdot, \cdot)$ to be slightly more general.

In the following, we outline the basics of the so-called *linearized stability* theory. The perturbation $w(t) := \tilde{u}(t) - \hat{u}$ satisfies the nonlinear *perturbation equation*

$$
\begin{aligned}
0 &= m(\partial_t w, \psi) + a(\hat{u}+w)(\psi) - a(\hat{u})(\psi) \\
&= m(\partial_t w, \psi) + a'(\hat{u})(w, \psi) + \int_0^1 a''(\hat{u}+sw)(w, w, \psi)s \, ds,
\end{aligned}
$$

for $t \geq 0$. Taking $\psi = w$, we obtain

$$\frac{1}{2}\frac{d}{dt}m(w, w) + a'(\hat{u})(w, w) = -\int_0^1 a''(\hat{u}+sw)(w, w, w)s \, ds.$$

Assuming now that the perturbation $w(t)$ is initially small such the the cubic term on the right can be neglected, the initial decay or growth of $w(t)$ is determined by the spectral properties of the sesquilinear form $a'(\hat{u})(\cdot, \cdot)$ obtained by linearizing $a(\hat{u})(\cdot)$ at the base solution. Particularly, if the nonsymmetric *stability eigenvalue problem*

$$a'(\hat{u})(u, \psi) = \lambda m(u, \psi) \quad \forall \psi \in V, \tag{7.23}$$

has an eigenvalue $\lambda^{crit}$ with negative real part, then any perturbation having initially a nontrivial component in the direction of an eigenfunction of $\lambda^{crit}$ will grow initially and may eventually blow up.

We want to solve this stability eigenvalue problem numerically by a Galerkin approximation which reads as follow:

$$a'(\hat{u}_h)(u_h, \psi_h) = \lambda_h m(u_h, \psi_h) \quad \forall \psi_h \in V_h. \tag{7.24}$$

Our goal is to estimate the error in the critical eigenvalue, $\lambda^{crit} - \lambda_h^{crit}$, i.e. the 'first' eigenvalue with possibly negative real part, in terms of the residuals corresponding to primal and dual equations. Note that this includes both the error due to approximating the base solution $\hat{u}$, as well as of the eigenvalue. To this end, we embed this situation into the general framework of variational equations laid out above. We introduce the spaces

$$\mathcal{V} := V \times V \times \mathbb{C}, \quad \mathcal{V}_h := V_h \times V_h \times \mathbb{C},$$

consisting of elements $U := \{\hat{u}, u, \lambda\}$, $\Psi = \{\hat{\psi}, \psi, \nu\}$ and $U_h := \{\hat{u}_h, u_h, \lambda_h\}$, $\Psi_h = \{\hat{\psi}_h, \psi_h, \nu\}$, respectively. Using the semilinear form

$$A(U)(\Psi) := a(\hat{u})(\hat{\psi}) + \lambda m(u, \psi) - a'(\hat{u})(u, \psi) + \bar{\nu}\{m(u, u) - 1\},$$

the continuous and discrete problems can be written in compact form as follows:

$$A(U)(\Psi) = 0 \quad \forall \Psi \in \mathcal{V}, \tag{7.25}$$
$$A(U_h)(\Psi_h) = 0 \quad \forall \Psi_h \in \mathcal{V}_h, \tag{7.26}$$

For controling the eigenvalue error, we work again with the functional

$$J(\Phi) := \mu m(\varphi, \varphi), \quad \Phi = \{\hat{\varphi}, \varphi, \mu\},$$

such that $J(U) = \lambda m(u, u) = \lambda$. Then, the corresponding continuous and discrete dual solutions $Z = \{\hat{z}, z, \pi\} \in \mathcal{V}$ and $Z_h = \{\hat{z}_h, z_h, \pi_h\} \in \mathcal{V}_h$ are determined by the problems

$$A'(U)(\Phi, Z) = J'(U)(\Phi) \quad \forall \Phi \in \mathcal{V}, \tag{7.27}$$
$$A'(U_h)(\Phi_h, Z_h) = J'(U_h)(\Phi_h) \quad \forall \Phi_h \in \mathcal{V}_h. \tag{7.28}$$

A detailed computation shows that these equations are solved by $Z = U^* := \{\hat{z}, u^*, \lambda\}$ and $Z_h = U_h^* := \{\hat{z}_h, u_h^*, \lambda_h\}$, respectively, where $u^*$ and $u_h^*$ are the (normalized) dual eigenfunctions as before, and the dual base solutions $\hat{z}$ and $\hat{z}_h$ are determined by the dual problems

$$a'(\hat{u})(\varphi, \hat{z}) = -a''(\hat{u})(\varphi, u, u^*) \quad \forall \varphi \in V, \tag{7.29}$$
$$a'(\hat{u}_h)(\varphi_h, \hat{z}_h) = -a''(\hat{u}_h)(\varphi_h, u_h, u_h^*) \quad \forall \varphi \in V. \tag{7.30}$$

The corresponding residuals of the approximate base solutions are

$$\hat{\rho}(\hat{u}_h)(\cdot) := -a(\hat{u}_h)(\cdot),$$
$$\hat{\rho}^*(\hat{u}_h, \hat{z}_h)(\cdot) := a''(\hat{u}_h)(\cdot, u_h, u_h^*) - a'(\hat{u}_h)(\cdot, \hat{z}_h),$$

and those of the eigenpair approximations,

$$\rho(\hat{u}_h, u_h, \lambda_h)(\cdot) := a'(\hat{u}_h)(u_h, \cdot) - \lambda_h \, m(u_h, \cdot),$$
$$\rho^*(\hat{u}_h, u_h^*, \lambda_h^*)(\cdot) := a'(\hat{u}_h)(\cdot, u_h^*) - \bar{\lambda}_h^* \, m(\cdot, u_h^*).$$

We collect the foregoing findings in the following proposition:

**Proposition 7.9.** *With the above residuals, we have the error representation*

$$\lambda - \lambda_h = \tfrac{1}{2}\hat{\rho}(\hat{u}_h)(\hat{z} - \hat{\psi}_h) + \tfrac{1}{2}\hat{\rho}^*(\hat{u}_h, \hat{z}_h)(\hat{u} - \hat{\varphi}_h)$$
$$+ \tfrac{1}{2}\rho(\hat{u}_h, u_h, \lambda_h)(u^* - \psi_h) + \tfrac{1}{2}\rho^*(\hat{u}_h, u_h^*, \lambda_h^*)(u - \varphi_h) + \mathcal{R}_h, \tag{7.31}$$

*for arbitrary* $\hat{\psi}_h, \psi_h, \hat{\varphi}_h, \varphi_h \in V_h$. *The remainder* $\mathcal{R}_h$ *is cubic in the errors* $\hat{e}^u := \hat{u} - \hat{u}_h$, $\hat{e}^z := \hat{z} - \hat{z}_h$, *and* $e^\lambda := \lambda - \lambda_h$, $e^u := u - u_h$, $e^{u*} := u^* - u_h^*$:

$$\mathcal{R}_h = \tfrac{1}{2}e^\lambda m(e^u, e^{u*}) + \tfrac{1}{2}\int_0^1 \Big\{ a''''(\hat{u}_h + s\hat{e}^u)(\hat{e}^u, \hat{e}^u, \hat{e}^u, u_h + se^u, u_h^* + se^{u*})$$
$$- a'''(\hat{u}_h + s\hat{e}^u)(\hat{e}^u, \hat{e}^u, \hat{e}^u, \hat{z}_h + s\hat{e}^z)$$
$$+ 3a'''(\hat{u}_h + s\hat{e}^u)(\hat{e}^u, \hat{e}^u, e^u, u_h^* + se^{u*})$$
$$+ 3a'''(\hat{u}_h + s\hat{e}^u)(\hat{e}^u, \hat{e}^u, u_h + se^u, e^{u*})$$
$$+ 6a''(\hat{u}_h + s\hat{e}^u)(\hat{e}^u, e^u, e^{u*})$$
$$- 3a''(\hat{u}_h + s\hat{e}^u)(\hat{e}^u, \hat{e}^u, \hat{e}^{u*}) \Big\} s(s-1)\, ds.$$

## Application to a model case

We want to apply this abstract result to the concrete situation of the nonsymmetric model problem from Example 6.1 (vector Burgers equation). Here, the stability eigenvalue problem has the form

$$-\Delta u + \hat{u} \cdot \nabla u + u \cdot \nabla \hat{u} = \lambda u, \quad \text{in } \Omega, \quad u_{|\partial\Omega} = 0. \tag{7.32}$$

Its finite element approximation computes the discrete base solution $\hat{u}_h$ by solving

$$(\nabla \hat{u}_h, \nabla \psi_h) + (\hat{u}_h \cdot \nabla \hat{u}_h, \psi_h) = (f, \psi_h) \quad \forall \psi_h \in V_h, \tag{7.33}$$

and then determines the corresponding eigenvalue $\lambda_h$ from the eigenvalue problem

$$(\nabla u_h, \nabla \psi_h) + (\hat{u}_h \cdot \nabla u_h + u_h \cdot \nabla \hat{u}_h, \psi_h) = \lambda_h (u_h, \psi_h) \quad \forall \psi_h \in V_h. \tag{7.34}$$

The corresponding discrete dual problem determining $\hat{z}_h \in V_h$ reads

$$(\nabla\varphi_h, \nabla\hat{z}_h) + (\varphi_h, \nabla\hat{u}_h\hat{z}_h - \nabla\cdot(\hat{u}_h \otimes \hat{z}_h))$$
$$= (\varphi_h, \nabla u_h u_h^* - \nabla\cdot(u_h \otimes u_h^*)) \quad \forall\varphi_h \in V_h, \tag{7.35}$$

and that determining the discrete dual eigenpair $\{u_h^*, \lambda_h^*\}$,

$$(\nabla\varphi_h, \nabla u_h^*) + (\varphi_h, \nabla\hat{u}_h u_h^* - \nabla\cdot(\hat{u}_h \otimes u_h^*)) = \bar{\lambda}_h^* (\varphi_h, u_h^*) \quad \forall\varphi_h \in V_h. \tag{7.36}$$

Then, the associated cell residuals are

$$\hat{R}_{h|K} := f + \Delta\hat{u}_h - \hat{u}_h\cdot\nabla\hat{u}_h,$$
$$\hat{R}_{h|K}^* := \Delta\hat{z}_h + \nabla u_h u_h^* - \nabla\cdot(u_h \otimes u_h^*) - \nabla\hat{u}_h\hat{z}_h + \nabla\cdot(\hat{u}_h \otimes \hat{z}_h),$$
$$R_{h|K} := -\Delta u_h + \hat{u}_h\cdot\nabla u_h + u_h\cdot\nabla\hat{u}_h - \lambda_h u_h,$$
$$R_{h|K}^* := -\Delta u_h^* - \nabla\cdot(\hat{u}_h \otimes u_h^*) + \nabla\hat{u}_h u_h^* - \bar{\lambda}_h^* u_h^*,$$

while the associated edge residuals $\hat{r}_{h|\Gamma}$, $\hat{r}_{h|\Gamma}^*$ and $r_{h|\Gamma}$, $r_{h|\Gamma}^*$ have the same form as above in the case of the simple Poisson equation. For this situation Proposition 7.9 yields the following result:

**Proposition 7.10.** *Using the notation from above, and assuming again that*

$$|m(u - u_h, u^* - u_h^*)| \leq 1,$$

*we have the a posterior error estimate*

$$|\lambda - \lambda_h| \leq \eta_\lambda^\omega := \sum_{K \in \mathbb{T}_h} \left\{ \hat{\rho}_K \hat{\omega}_K^* + \hat{\rho}_K^* \hat{\omega}_K + \rho_K \omega_K^* + \rho_K^* \omega_K \right\} + \mathcal{R}_h. \tag{7.37}$$

*The cell residuals and weights are defined by*

$$\hat{\rho}_K := \left( \|\hat{R}_h\|_K^2 + h_K^{-1/2}\|\hat{r}_h\|_{\partial K}^2 \right)^{1/2},$$
$$\hat{\omega}_K^* := \left( \|\hat{z} - \hat{\psi}_h\|_K^2 + h_K^{1/2}\|\hat{z} - \hat{\psi}_h\|_{\partial K}^2 \right)^{1/2},$$
$$\hat{\rho}_K^* := \left( \|\hat{R}_h^*\|_K^2 + h_K^{-1/2}\|\hat{r}_h^*\|_{\partial K}^2 \right)^{1/2},$$
$$\hat{\omega}_K := \left( \|\hat{u} - \hat{\varphi}_h\|_K^2 + \tfrac{1}{2}h_K^{1/2}\|\hat{u} - \hat{\varphi}_h\|_{\partial K}^2 \right)^{1/2},$$
$$\rho_K := \left( \|R_h\|_K^2 + h_K^{-1/2}\|r_h\|_{\partial K}^2 \right)^{1/2},$$
$$\omega_K^* := \left( \|u^* - \psi_h\|_K^2 + h_K^{1/2}\|u^* - \psi_h\|_{\partial K}^2 \right)^{1/2},$$
$$\rho_K^* := \left( \|R_h^*\|_K^2 + h_K^{-1/2}\|r_h^*\|_{\partial K}^2 \right)^{1/2},$$
$$\omega_K := \left( \|u - \varphi_h\|_K^2 + \tfrac{1}{2}h_K^{1/2}\|u - \varphi_h\|_{\partial K}^2 \right)^{1/2}.$$

*for arbitrary* $\hat{\psi}_h, \psi_h, \hat{\varphi}_h, \varphi_h \in V_h$, *and the remainder* $\hat{\mathcal{R}}_h$ *is cubic in the errors* $\hat{e}^u := \hat{u} - \hat{u}_h$ *and* $\hat{e}^{u*} := \hat{u}^* - \hat{u}_h^*$ :

$$\mathcal{R}_h = -\left( \nabla(\hat{e}^u \otimes e^u - \tfrac{1}{2}\hat{e}^u \hat{\otimes} e^u), e^{u*} \right).$$

An error estimator as derived in Proposition 7.10 will be applied below in Chapter 11 to the approximation of the stability eigenvalue problem of the Navier-Stokes equations. There, we will demonstrate the interplay of the different components in the error estimator $\eta_\lambda^\omega$.

## 7.4  Exercises

*Exercise 7.1.* For the Galerkin approximation of a symmetric eigenvalue problem

$$a(u, \psi) = \lambda\,(u, \psi) \quad \forall \psi \in V,$$

with a symmetric bilinear form $a(\cdot, \cdot)$ on a (real) Hilbert space $V$, the general a posteriori error representation reduces to the form

$$\lambda - \lambda_h = \rho(u_h, \lambda_h)(u - \psi_h) + \tfrac{1}{2}(\lambda - \lambda_h)\|u - u_h\|^2, \quad \psi \in V_h.$$

Derive this identity by direct algebraic manipulation.

*Exercise 7.2.* Consider the model convection-diffusion eigenvalue problem

$$-\Delta v + b\cdot\nabla v \;=\; \lambda v \quad \text{in } \Omega, \quad v_{|\partial\Omega} = 0.$$

Under the assumption that the eigenfunctions have $H^2$-regularity, and that

$$|(u - u_h, u^* - u_h^*)| \leq 1,$$

prove the a posteriori error bound

$$|\lambda - \lambda_h| \leq \eta_\lambda^{(2)} := c_\lambda \Big( \sum_{K \in \mathbb{T}_h} h_K^4 \{\rho_K^2 + \rho_K^{*2}\} \Big)^{1/2},$$

with a constant $c_\lambda = \mathcal{O}(|\lambda|)$.

*Exercise 7.3.* Consider the $d$-dimensional Burgers equation

$$-\nu\Delta u + u\cdot\nabla u = f \quad \text{in } \Omega, \quad u_{|\partial\Omega} = 0,$$

for a vector function $u : \Omega \to \mathbb{R}^d$. Suppose that $\hat{u} \in V := H_0^1(\Omega)^d$ is a solution. Show that if all eigenvalues of the *symmetric* stability eigenvalue problem

$$-\nu\Delta w + \tfrac{1}{2}\{\nabla\hat{u} + \nabla\hat{u}^T - \nabla\cdot\hat{u}I\}w = \lambda w$$

are positive, then $\hat{u}$ is *dynamically stable*. This criterion for stability is much stronger than that of *linearized stability* since it allows perturbations of any size and also applies to nonstationary solutions. Therefore, it cannot be expected to be satisfied in many practically interesting situations.

*Exercise* 7.4 *(Practical exercise).* Consider the nonlinear boundary value problem
of Exercise 5.3,

$$-\Delta u - u^3 = f, \quad \text{in } \Omega, \quad u_{|\partial\Omega} = 0,$$

where $\Omega$ is the domain defined in Exercise 3.3. The discretization is by the usual
bilinear finite elements. For $f \equiv \alpha = 1, \dots, 75$, compute the solution on increas-
ingly refined meshes and investigate its dynamic stability using the criterion of
*linearized stability*. Monitor the behavior of the eigenvalue error estimator and, in
particular, its components measuring the approximation of the base solution and
that of the eigenvalue. Explain the observed results.

# Chapter 8

# Optimization Problems

As another important application for the general theory of the DWR method developed in Chapter 6, we consider optimization problems with PDE constraints as discussed in Kapp [96] and Becker et al. [28]. In abstract variational notation, such problems are posed in a *state space* $V$ and a *control space* $Q$, on which state and control operators, associated with the forms $A(\cdot)(\cdot)$ and $B(\cdot,\cdot)$, respectively, act and on which the *cost functional* $J(\cdot\cdot)$ is defined. The problem then reads: *Minimize $J(u,q)$ for pairs $\{u,q\} \in V \times Q$, subject to the constraint*

$$A(u)(\psi) + B(q,\psi) = 0 \quad \forall \psi \in V. \tag{8.1}$$

Their Galerkin approximation uses subspaces $V_h \times Q_h \subset V \times Q$ and reads as follows: *Minimize $J(u_h, q_h)$ for pairs $\{u_h, q_h\} \in V_h \times Q_h$, subject to the constraint*

$$A(u_h)(\psi_h) + B(q_h,\psi_h) = 0 \quad \forall \psi_h \in V_h. \tag{8.2}$$

As before, the semilinear form $A(\cdot)(\cdot)$ is assumed to be sufficiently often differentiable, while for simplicity, the cost functional $J(\cdot,\cdot)$ is assumed as linear or quadratic and the control form $B(\cdot,\cdot)$ as linear. This abstract setting is illustrated by the model situation described in the following example.

*Example* 8.1. We consider the state equation

$$-\Delta u + s(u) = f \quad \text{in } \Omega, \tag{8.3}$$

posed on the $T$-shaped domain $\Omega \subset \mathbb{R}^2$ shown in Figure 8.1, with the Neumann boundary conditions

$$\partial_n u = q \text{ on } \Gamma_C, \quad \partial_n u = 0 \text{ on } \partial\Omega \backslash \Gamma_C.$$

Here, the control variable $q$ prescribed along the *control boundary* $\Gamma_C$ is to be adjusted to minimize the cost functional

$$J(u,q) = \tfrac{1}{2}\|u - u_0\|_{\Gamma_O}^2 + \tfrac{1}{2}\alpha\|q\|_{\Gamma_C}^2.$$

The cost functional $J(u, q)$ measures the derivation of the Dirichlet values of the solution $u$ along the *observation boundary* $\Gamma_O$, the *observations*, from the prescribed function $u_0$. The parameter $\alpha \geq 0$ expresses the amount of regularization in the cost functional. In this problem, the generic solution spaces are $V := H^1(\Omega)$ for the state variable $u$, and $Q := L^2(\Gamma_C)$ for the control variable $q$. The corresponding variational formulation is

$$(\nabla u, \nabla \psi) + (s(u), \psi) - (q, \psi)_{\Gamma_C} = (f, \psi) \quad \forall \psi \in V, \tag{8.4}$$

with the *control form* $B(q, \psi) := -(q, \psi)_{\Gamma_C}$.

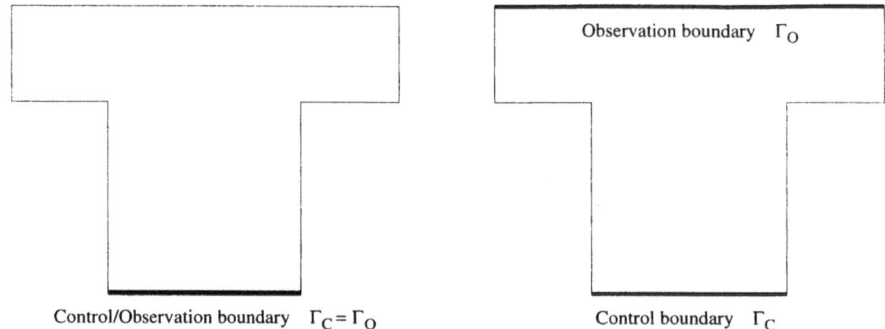

Figure 8.1: *Configuration of the boundary control model problem on a T-domain: configuration 1 (left), configuration 2 (right).*

The economical numerical treatment of optimization problems with PDE constraints requires some preliminary thoughts:

- *What is our notion of admissible states* $u = u(q)$ *?*
  In general, under discretization the state equation cannot be satisfied exactly, but only in an approximate sense. Since achieving accuracy in the approximation of PDEs is 'expensive', the amount of *admissibility* of the state which is relevant for the optimization process becomes a crucial question.

- *How to 'measure' admissibility?*
  In solving ODEs, we may require the error to be uniformly 'small'. However, in the context of PDEs, the choice of an adequate error measure is not clear. Possible candidates such as the 'global' energy-norm and $L^2$ norm or alternatively more 'localized' error measures may lead to rather different degrees of admissibility of the state variable.

# 8.1 A posteriori error analysis via Lagrange formalism

In order to develop the adaptive discretization of optimal control problems as introduced above, we embed them into the abstract framework laid out in Chapter 6. This means that we use the so-called *indirect method of optimization* in which stationary points of the associated Lagrangian functional are computed among which the possible local minima can be found. The more traditional *direct method of optimization* tries to minimize the cost functional by working only on the set of admissible functions, i.e. those pairs $\{u, q\}$ satisfying the state equation.

In this case, we define the Lagrangian functional by

$$\mathcal{L}(u, q, \lambda) := J(u, q) - A(u)(\lambda) - B(q, \lambda),$$

with the *primal variable* $u \in V$, the *control variable* $q \in Q$, and the *adjoint variable* $\lambda \in V$ (Lagrangian multiplier). Again, we seek stationary points $x := \{u, q, \lambda\} \in X := V \times Q \times V$ of $\mathcal{L}(\cdot, \cdot, \cdot)$ which are determined by the following system of equations (*Euler-Lagrange system* or *optimality system*):

$$J_u'(u, q)(\varphi) - A'(u)(\varphi, \lambda) = 0 \quad \forall \varphi \in V, \tag{8.5}$$

$$J_q'(u, q)(\chi) - B(\chi, \lambda) = 0 \quad \forall \chi \in Q, \tag{8.6}$$

$$-A(u)(\psi) - B(q, \psi) = 0 \quad \forall \psi \in V. \tag{8.7}$$

Notice that the third equation is just the state equation to be satisfied by any admissible pair $\{u, q\}$. The Galerkin approximation determines $x_h := \{u_h, q_h, \lambda_h\} \in X_h := V_h \times Q_h \times V_h$ by a corresponding system of discrete equations:

$$J_u'(u_h, q_h)(\varphi_h) - A'(u_h)(\varphi_h, \lambda_h) = 0 \quad \forall \varphi_h \in V_h, \tag{8.8}$$

$$J_q'(u_h, q_h)(\chi_h) - B(\chi_h, \lambda_h) = 0 \quad \forall \chi_h \in Q_h, \tag{8.9}$$

$$-A(u_h)(\psi_h) - B(q_h, \psi_h) = 0 \quad \forall \psi_h \in V_h. \tag{8.10}$$

We assume that both systems (8.5)–(8.7) and (8.8)–(8.10) have solutions. This has to be shown in a concrete situations using the particular structural properties of the problem considered.

It appears quite natural to control the error of this approximation with respect to the cost functional, i.e. by estimating $J(u, q) - J(u_h, q_h)$ in terms of the *dual*, *control*, and *primal* residuals defined by

$$\rho^*(u_h, q_h, \lambda_h)(\cdot) := J_u'(u_h, q_h)(\cdot) - A'(u_h)(\cdot, \lambda_h),$$

$$\rho^q(u_h, q_h, \lambda_h)(\cdot) := J_q'(u_h, q_h)(\cdot) - B(\cdot, \lambda_h),$$

$$\rho(u_h, q_h)(\cdot) := -A(u_h)(\cdot) - B(\cdot, q_h).$$

To this situation we can directly apply the abstract result of Proposition 6.1:

**Proposition 8.1.** *For the approximation of the Euler-Lagrange system (8.5)–(8.7) by the system (8.8)–(8.10), we have the error representation*

$$J(u,q) - J(u_h, q_h) = \tfrac{1}{2}\rho^*(u_h, q_h, \lambda_h)(u - \varphi_h) + \tfrac{1}{2}\rho^q(u_h, q_h, \lambda_h)(q - \chi_h)$$
$$+ \tfrac{1}{2}\rho(u_h, q_h)(\lambda - \psi_h) + \mathcal{R}_h, \tag{8.11}$$

*for arbitrary $\varphi_h$, $\psi_h \in V_h$ and $\chi_h \in Q_h$. The remainder $\mathcal{R}_h$ is cubic in the errors $e^u := u - u_h$, $e^q := q - q_h$, $e^\lambda := \lambda - \lambda_h$.*

*Proof.* For the proof, we recall the abstract error representation from Proposition 6.1, which applied to the Lagrangian functional reads as follows:

$$\mathcal{L}(x) - \mathcal{L}(x_h) = \tfrac{1}{2}\mathcal{L}'(x_h)(x - y_h) + \mathcal{R}_h, \quad y_h \in X_h, \tag{8.12}$$

with a remainder term $\mathcal{R}_h$ which is cubic in the error $e^x := x - x_h$,

$$\mathcal{R}_h := \tfrac{1}{2}\int_0^1 \mathcal{L}'''(x_h + se^x)(e^x, e^x, e^x)\, s(s-1)\, ds,$$

where $x = \{u, q, \lambda\}$ and $x_h = \{u_h, q_h, \lambda_h\}$. By computing the explicit form of the derivative $\mathcal{L}'$ and noting that

$$\mathcal{L}(x) - \mathcal{L}(x_h) = J(u,q) - J(u_h, q_h),$$

we obtain the residual part of the asserted representation. Since the control form $B(\cdot, \cdot)$ is assumed to be linear, $\mathcal{L}(u, q, \lambda)$ is linear in $\lambda$, and $J(u, q)$ is assumed to be at most quadratic, the third derivative of $\mathcal{L}(\cdot)$ consists of only two terms, namely,

$$-A'''(u_h + se^u)(e^u, e^u, e^u, \lambda_h + se^\lambda) - 3A''(u_h + se^u)(e^u, e^u, e^\lambda).$$

This completes the proof.  □

*Remark 8.2.* We note that for the concrete situation of equation (8.3) the remainder term in the error representation (8.11) has the form

$$\mathcal{R}_h = -\tfrac{1}{2}\int_0^1 \left\{ (s'''(u_h + se^u)e^{u3}, \lambda_h + se^\lambda) + 3(s''(u_h + se^u)e^{u2}, e^\lambda) \right\} s(s-1)\, ds.$$

*Remark 8.3.* In the a posteriori error representation (8.11) only the primary variables $x = \{u, q, \lambda\}$ occur which have to be computed simultaneously in the indirect method of optimization. Hence, no extra dual problem has to be solved which makes this approach of error estimation with respect to the natural cost functional especially attractive. The mechanism underlying this property is the same which was already seen in the context of eigenvalue computation in Proposition 7.1 and earlier in Chapter 3 in deriving bounds for the error in the energy-norm. In all these examples, the error is to be controlled with respect to the functional from which the variational equations are derived as first-order stationarity conditions.

*Remark* 8.4. For the practical solution of the system (8.8)–(8.10), we use a nested iteration consisting of an outer Newton iteration with automatic mesh adaptation and an inner linear multigrid iteration; see Section 6.2. Since this process starts from a coarse mesh which is then successively refined, we may also speak of 'model enrichment' in adaptively solving the optimality system.

*Remark* 8.5. The indirect method of optimization combined with discretization yields approximations $\{u_h, q_h, \lambda_h\}$ which may be admissible only in a very weak sense. Nevertheless, by construction, on the optimally adapted mesh $\mathbb{T}_h$ the obtained control $q_h^{\mathrm{opt}}$ yields a value of the cost functional $J(u_h^{\mathrm{opt}}, q_h^{\mathrm{opt}})$ which differs from the exact optimal value $J(u, q)$ only by the prescribed tolerance TOL. If for certain reasons a more admissible state variable $\tilde{u}_h$ is needed, then we may generate this in a post-processing step by solving the state equation on a finer mesh with the previously computed $q_h^{\mathrm{opt}}$ as fixed data:

$$A(\tilde{u}_h)(\varphi_h) = -B(q_h^{\mathrm{opt}}, \varphi_h) \quad \forall \varphi_h \in \tilde{V}_h.$$

## 8.2 Application to a boundary control problem

We consider the example described above with the state equation (8.1) and non-linearity $s(u) := u^3 - u$, posed on the $T$-domain shown in Figure 8.1. Then, the Euler-Lagrange system reads as follows:

$$
\begin{aligned}
(\varphi, u - u_0)_{\Gamma_O} - (\nabla\varphi, \nabla\lambda)_\Omega - (s'(u)\varphi, \lambda)_\Omega &= 0 \quad &\forall \varphi \in V, \\
\alpha(q, \chi)_{\Gamma_C} + (\lambda, \chi)_{\Gamma_C} &= 0 \quad &\forall \chi \in Q, \\
-(\nabla u, \nabla\psi)_\Omega - (s(u), \psi)_\Omega + (f, \psi)_\Omega + (q, \psi)_{\Gamma_C} &= 0 \quad &\forall \psi \in V.
\end{aligned}
\tag{8.13}
$$

For illustration, we also state the strong form of this system:

$$
\begin{aligned}
-\Delta\lambda + s'(u)\lambda &= 0 \ \text{ in } \Omega, \quad \partial_n\lambda_{|\Gamma_O} = u - u_0, \ \partial_n\lambda_{|\partial\Omega\setminus\Gamma_O} = 0, \\
\lambda_{|\Gamma_C} &= -\alpha q_{|\Gamma_C}, \\
-\Delta u + s(u) &= f \ \text{ in } \Omega, \quad \partial_n u_{|\Gamma_C} = q, \ \partial_n u_{|\partial\Omega\setminus\Gamma_C} = 0.
\end{aligned}
\tag{8.14}
$$

For the Galerkin approximation of this optimality system, we use the standard spaces $V_h$ of bilinear finite elements on meshes $\mathbb{T}_h$ for the primal and adjoint variables $u_h$ and $\lambda_h$, respectively. For the discretization of the control variable $q$, we use the space of traces of normal derivatives of functions in $V_h$ on $\Gamma_C$, i.e. piecewise linear shape functions. The discrete *optimality system* reads

$$
\begin{aligned}
(\varphi_h, u_h - u_0)_{\Gamma_O} - (\nabla\varphi_h, \nabla\lambda_h)_\Omega - (s'(u_h)\varphi_h, \lambda_h)_\Omega &= 0 \quad &\forall \varphi_h \in V_h, \\
\alpha(q_h, \chi_h)_{\Gamma_C} + (\lambda_h, \chi_h)_{\Gamma_C} &= 0 \quad &\forall \chi_h \in Q_h, \\
-(\nabla u_h, \nabla\psi_h)_\Omega - (s(u_h), \psi_h)_\Omega + (f, \psi_h)_\Omega + (q_h, \psi_h)_{\Gamma_C} &= 0 \quad &\forall \psi_h \in V_h.
\end{aligned}
\tag{8.15}
$$

For steering the mesh adaptation, we use the following a posteriori error estimator
which is obtained from Proposition 8.1:

$$|J(u,q) - J(u_h, q_h)| \approx \eta_\omega := \tfrac{1}{2} \sum_{K \in \mathbb{T}_h} \{ \rho_K^u \, \omega_K^\lambda + \rho_K^q \, \omega_K^q + \rho_K^\lambda \, \omega_K^u \}, \qquad (8.16)$$

where the residuals and weights are for $x_h = \{u_h, q_h, \lambda_h\}$ defined by

$$\rho_K^\lambda := \|R_h^\lambda\|_K + h_K^{-1/2}\|r_h^\lambda\|_{\partial K}, \qquad \omega_K^u := \|u - I_h u\|_K + h_K^{1/2}\|u - I_h u\|_{\partial K},$$

$$\rho_K^q := h_K^{-1/2}\|r_h^q\|_{\partial K}, \qquad\qquad\qquad \omega_K^q := h_K^{1/2}\|q - I_h q\|_{\partial K},$$

$$\rho_K^u := \|R_h^u\|_K + h_K^{-1/2}\|r_h^u\|_{\partial K}, \qquad \omega_K^\lambda := \|\lambda - I_h \lambda\|_K + h_K^{1/2}\|\lambda - I_h \lambda\|_{\partial K}.$$

with suitable approximations $\{I_h u, I_h q, I_h \lambda\} \in V_h \times Q_h \times V_h$. The cell residuals
are given by

$$R_{h|K}^u := f + \Delta u_h - s(u_h), \qquad R_{h|K}^\lambda := \Delta \lambda_h - s'(u_h)\lambda_h,$$

and edge residuals by

$$r_{h|\Gamma}^u := \begin{cases} \tfrac{1}{2}[\partial_n u_h], & \text{if } \Gamma \not\subset \partial\Omega, \\ \partial_n u_h, & \text{if } \Gamma \subset \partial\Omega \backslash \Gamma_C, \\ \partial_n u_h - q_h, & \text{if } \Gamma \subset \Gamma_C, \end{cases} \qquad r_{h|\Gamma}^q := \begin{cases} \lambda_h + \alpha q_h, & \text{if } \Gamma \subset \Gamma_C, \\ 0, & \text{if } \Gamma \not\subset \Gamma_C. \end{cases}$$

$$r_{h|\Gamma}^\lambda := \begin{cases} \tfrac{1}{2}[\partial_n \lambda_h], & \text{if } \Gamma \not\subset \partial\Omega, \\ \partial_n \lambda_h, & \text{if } \Gamma \subset \partial\Omega \backslash \Gamma_O, \\ \partial_n \lambda_h - u_h + u_0, & \text{if } \Gamma \subset \Gamma_O. \end{cases}$$

We will compare the performance of the weighted error estimator (8.16) with
more traditional error estimators. Control of the error in the energy-norm of the
state equation alone leads us to the 'reduced' a posteriori error estimator

$$\eta_E^{\text{red}} := c_I \Big( \sum_{K \in \mathbb{T}_h} h_K^2 \, (\rho_K^u)^2 \Big)^{1/2}, \qquad (8.17)$$

while taking into account the full optimality system (8.5)–(8.7) to

$$\eta_E := c_I \Big( \sum_{K \in \mathbb{T}_h} h_K^2 \, \{ (\rho_K^u)^2 + (\rho_K^\lambda)^2 + (\rho_K^q)^2 \} \Big)^{1/2}, \qquad (8.18)$$

with the residual terms as defined above and interpolation constants $c_I$ usually
set to be one. These *ad hoc* criteria aim at satisfying the state equation or the
whole set of optimality conditions uniformly with good accuracy. However, this
concept seems questionable since it does not take into account the sensitivity of the
cost functional with respect to the discretization in different parts of the domain.
Capturing these dependencies is the particular feature of the DWR method.

**Numerical results for Configuration 1**

The configuration is as shown in Figure 8.1, with $u_0 = \sin(0.19x)$ and $\alpha = 0$. This test case represents an extreme situation since here the observation $u_{|\Gamma_O}$ is evaluated right at the control boundary $\Gamma_C$, i.e., the flow of information does not have to pass through the domain. Hence, the corner singularities do not much affect the optimization process and should not induce mesh refinement.

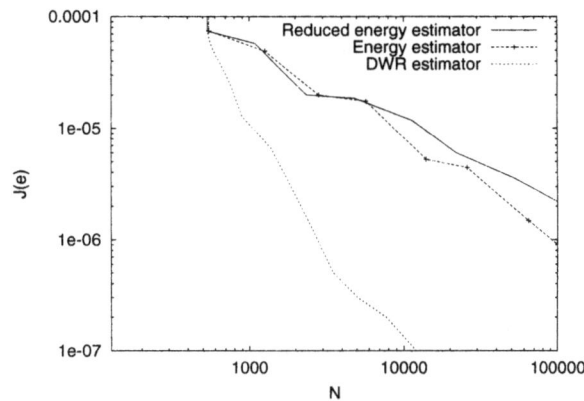

| N | $E_{\mathrm{rel}}$ | $I_{\mathrm{eff}}$ |
|---|---|---|
| 596 | $2.5\cdot10^{-4}$ | 2.94 |
| 1616 | $2.3\cdot10^{-4}$ | 1.24 |
| 5084 | $8.2\cdot10^{-5}$ | 2.17 |
| 8648 | $4.2\cdot10^{-5}$ | 3.44 |
| 15512 | $3.9\cdot10^{-5}$ | 2.32 |

Figure 8.2: *Relative error $E_{\mathrm{rel}}$ and effectivity index $I_{\mathrm{eff}}$ obtained by the error estimator $\eta_\omega$ (left), and efficiency of the meshes generated by the estimators $\eta_E^{\mathrm{red}}$ (solid line), $\eta_E$ (dotted line, +), and $\eta_\omega$ (dotted line); from Kapp [96].*

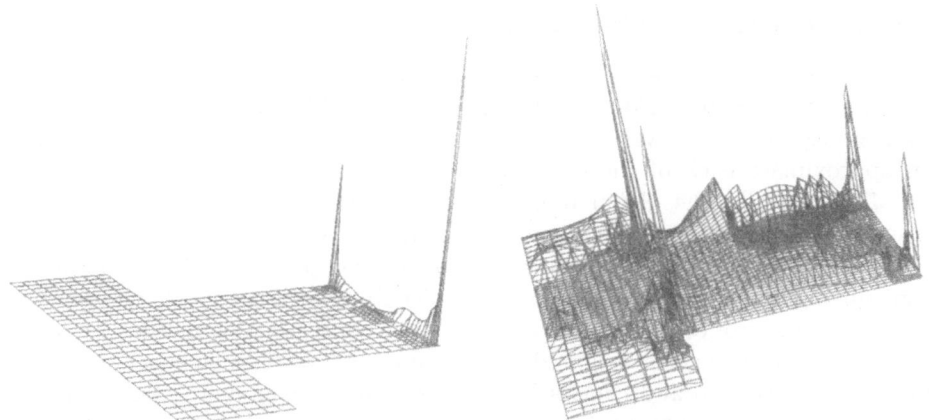

Figure 8.3: *Size of cell-error indicators $\eta_K$ in the weighted error estimator $\eta_\omega$ (left) and the energy-norm error estimator $\eta_E^{\mathrm{red}}$ (right); from Kapp [96].*

In Figure 8.2, we compare the efficiency of the meshes generated by the weighted error estimator and the energy-norm error indicators. The first one yields significantly more economical meshes; the optimal value ( $J(u_h, q_h) = 0.011948$ ) of the cost function is obtained with only $3,500$ cells compared to the $100,000$ cells needed by the energy-norm error estimator. Figure 8.3 shows the distribution of cell-error indicators derived from the different a posteriori error estimators, and Figure 8.4 shows the discrete solutions obtained on the corresponding adapted meshes.

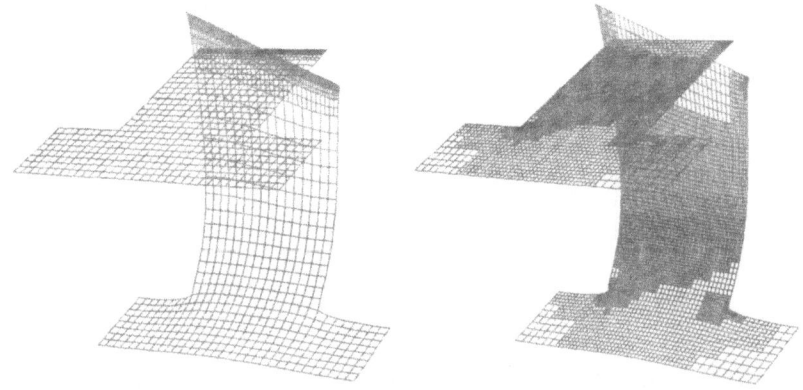

Figure 8.4: *Comparison of discrete solutions on adapted meshes obtained by the error estimators $\eta_\omega$ (left) and $\eta_E^{\text{red}}$ (right); from Kapp [96].*

**Numerical results for Configuration 2**

For the second test, we take the observations as $u_0 \equiv 1$ and set the regularization parameter to $\alpha = 0.1$. Now, there exist several stationary points of the Euler-Lagrange equations. By varying the starting values for the Newton iteration, we can approximate each of these solutions. The trivial solution corresponding to $u \equiv 1$ is actually the global minimum of the cost functional. The two other computed solutions correspond to a local minimum and a local maximum.

The effectivity of the error estimator $\eta_\omega$ for computing the local minimum is shown in Figure 8.5. Figure 8.6 shows the distribution of the local cell indicators $\eta_K$ for the two error estimators $\eta_\omega$ and $\eta_E^{\text{red}}$; the corresponding meshes are seen in Figure 8.7. Obviously, the weighted error estimator induces a stronger refinement along the observation and control boundaries which seems more relevant for the optimization process than resolving the corner singularities. However, in this case the error contribution by the interior cells around the reentrant corners is dominant over that by the boundary cells, such that also the energy-norm error estimator yields reasonable meshes for the optimization process. This explains why the gain in efficiency (about 25%) of the error estimator $\eta_\omega$ over $\eta_E^{\text{red}}$ is less significant here compared to the previous example.

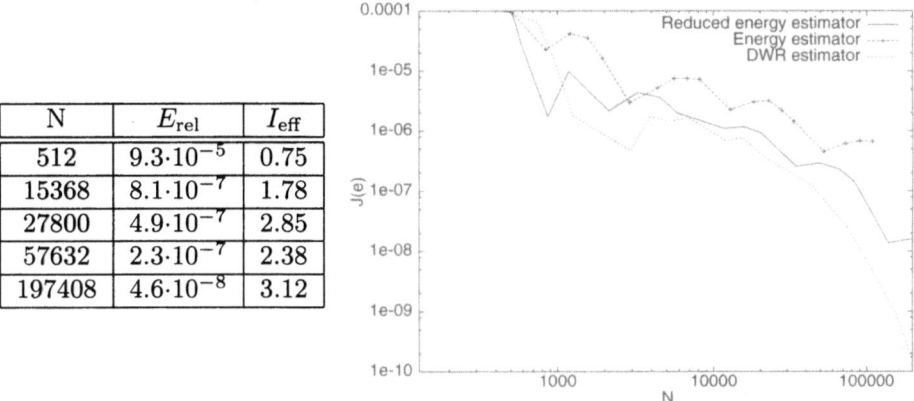

| N | $E_{\text{rel}}$ | $I_{\text{eff}}$ |
|---|---|---|
| 512 | $9.3 \cdot 10^{-5}$ | 0.75 |
| 15368 | $8.1 \cdot 10^{-7}$ | 1.78 |
| 27800 | $4.9 \cdot 10^{-7}$ | 2.85 |
| 57632 | $2.3 \cdot 10^{-7}$ | 2.38 |
| 197408 | $4.6 \cdot 10^{-8}$ | 3.12 |

Figure 8.5: *Relative error $E_{\text{rel}}$ and effectivity index $I_{\text{eff}}$ obtained by the error estimator $\eta_\omega$ (left), and efficiency of the meshes generated by the estimators $\eta_E^{\text{red}}$ (solid line), $\eta_E$ (dashed line, +), and $\eta_\omega$ (dotted line); from Kapp [96].*

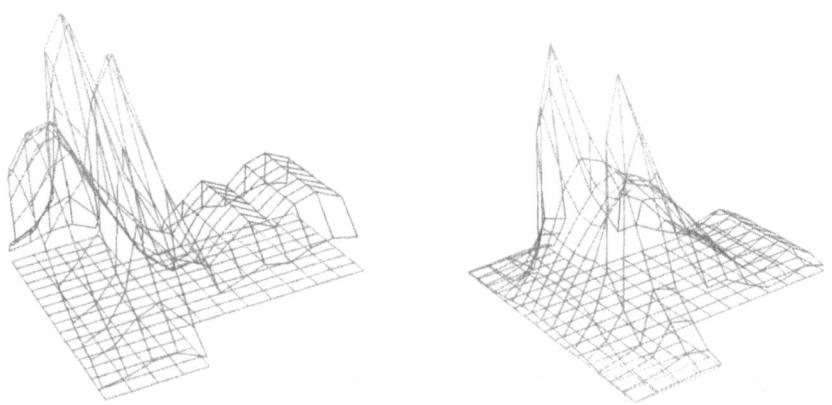

Figure 8.6: *Size of cell-error indicators $\eta_K$ in the weighted error estimator $\eta_\omega$ (left) and the energy-norm error estimator $\eta_E^{\text{red}}$ (right); from Kapp [96].*

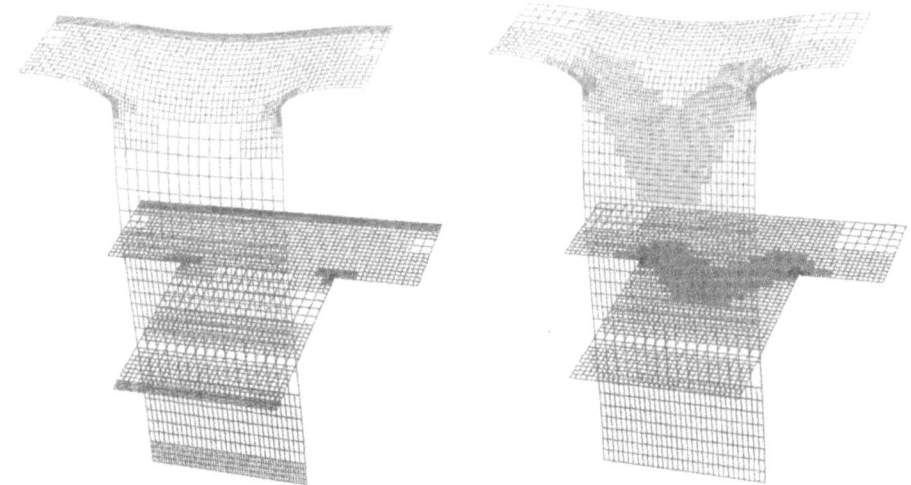

Figure 8.7: *Comparison of discrete solutions on adapted meshes obtained by the error estimators $\eta_\omega$ (left) and $\eta_E^{\mathrm{red}}$ (right); from Kapp [96].*

## 8.3   Application to parameter estimation

Another important class of optimization problems related to PDEs are so-called *parameter estimation problems*. Here, we consider the model problem

$$-\Delta u + qu = f \ \text{ in } \Omega, \quad u_{|\partial\Omega} = 0. \tag{8.19}$$

The goal is to determine the coefficient $q$ by comparing the resulting state $u(q)$ with given measurements $\bar{u}$,

$$J(u, q) := \tfrac{1}{2}\|u - \bar{u}\|^2 + \tfrac{1}{2}\alpha\|q\|^2 \to \min,$$

where $J(\cdot, \cdot)$ is the *misfit functional*, and $\alpha > 0$ is a small regularization parameter. Reformulating the problem within the Euler-Lagrange approach, we seek a stationary point $\{u, q, \lambda\}$ of the Lagrangian functional

$$L(u, q, \lambda) := J(u, q) - (\nabla u, \nabla \lambda) - (qu, \lambda) + (f, \lambda),$$

which is determined by the optimality system

$$-(\varphi, u - \bar{u}) + (\nabla\varphi, \nabla\lambda) + (q\varphi, \lambda) = 0 \quad \forall\varphi, \tag{8.20}$$

$$\alpha(\chi, q) - (\chi u, \lambda) = 0 \quad \forall\chi, \tag{8.21}$$

$$(\nabla u, \nabla\psi) + (qu, \psi) - (f, \psi) = 0 \quad \forall\psi. \tag{8.22}$$

Discretization by the usual finite element method yields $\{u_h, q_h, \lambda_h\}$, for which the following error representation holds:

$$J(u,q) - J(u_h, q_h) = \tfrac{1}{2}\rho^*(u_h, q_h, \lambda_h)(u - \varphi_h) + \tfrac{1}{2}\rho^q(u_h, q_h, \lambda_h)(q - \chi_h)$$
$$+ \tfrac{1}{2}\rho(u_h, q_h)(\lambda - \psi_h) + \mathcal{R}_h, \tag{8.23}$$

with residuals defined in the obvious way. Usually the numerical solution of the system (8.20)–(8.20) is difficult since it is ill-posed. This is especially the case when the parameter $q$ is not fully determined by the prescribed observations. Then, the error identity based on the 'artificial' cost functional $J(u,q)$ may be useless for steering the mesh adaptation. Indeed, if the parameter $q \geq 0$ is identifyable, we have $u = \bar{u}$ at the stationary point, and the adjoint variable $\lambda$ satisfies

$$-\Delta\lambda + q\lambda = u - \bar{u} = 0,$$

which implies that $\lambda \equiv 0$. This complication can be dealt with in several ways:

- An energy-norm-type a posteriori error estimate for $\alpha|q - q_h|^2$ can be derived based on a coercivity estimate for the saddle-point problem (8.20)–(8.22), see, e.g., Liu and Yan [103, 104]. However, the stability constant in this estimate is usually unknown and depends very unfavorably on the regularization parameter $\alpha$ rendering such a theoretical result almost useless.

- An a posteriori error estimate for a suitable norm of $q - q_h$ can be derived by an 'outer' duality argument similar to the approach used in Section 7.2 for estimating the error in functionals of eigenfunctions. In fact, an eigenvalue problem may be considered as a special type of parameter estimation problem; see Exercise 8.2. Exploiting the particular structure of the dual problem, this approach can be made computationally feasible also for parameter estimation problems, see [15] and Becker and Vexler [32].

## 8.4 Exercises

*Exercise* 8.1. Develop the detailed form of the general a posteriori error estimator applied to the optimal control problem

$$J(u) = \tfrac{1}{2}\int_\Omega |u - \bar{u}|^2\,dx \;\longrightarrow\; \min_{q\in\mathbb{R}},$$

with the PDE constraint

$$-\Delta u + qu = f \;\text{ in }\; \Omega, \quad u_{|\partial\Omega} = 0,$$

on a polygonal domain $\Omega \subset \mathbb{R}^2$, for a given function $\bar{u}$.

*Exercise* 8.2. The solution of the elliptic eigenvalue problem

$$-\Delta u = \lambda u \ \text{ in } \Omega, \quad u_{|\partial\Omega} = 0,$$

can be reformulated in the context of parameter estimation as follows:
Minimize

$$J(u) := \tfrac{1}{2}(u-1,1)^2 \ \rightarrow \ 0,$$

for pairs $\{u, \lambda\} \in H_0^1(\Omega) \times \mathbb{R}$ satisfying

$$(\nabla u, \nabla \psi) = \lambda(u, \psi) \quad \forall \psi \in H_0^1(\Omega).$$

Following the Euler-Lagrange approach, formulate the first-order necessary condition for computing the first (simple) eigenvalue $\lambda_{\min} > 0$ and show that in this case the adjoint solution vanishes, i.e., $z \equiv 0$. This shows that for guiding mesh refinement in computing eigenvalues another approach has to be used. We remark that, here, the cost functional $J(u)$ contains the *linear* normalization condition $(u, 1) = 1$ which is an alternative to the usual *quadratic* normalization condition $\|u\|^2 = 1$.

*Exercise* 8.3 *(Practical exercise)*. Consider the following optimization problem: try to choose the scalar heat flux parameter $q \in \mathbb{R}$ such that the solution of the stationary heat conduction problem

$$-\Delta u = 0 \ \text{ in } \Omega,$$
$$-\partial_n u = q \ \text{ on } \Gamma_C, \quad -\partial_n u = 0 \ \text{ on } \Gamma_N, \quad u = 0 \ \text{ on } \Gamma_D,$$

minimizes the misfit functional on the observation boundary $\Gamma_O$,

$$J(u) = \tfrac{1}{2}\|u - u_0\|_{\Gamma_O}^2,$$

with a prescribed target temperature $u_0 \equiv 1$. Here, $\Omega$ is the domain of Exercise 3.3, the control boundary is $\Gamma_C = \partial\Omega \cap \{y = -1\}$, and $\Gamma_N = \partial\Omega \cap \{x = \pm 1\}$, $\Gamma_D = \partial\Omega \backslash (\Gamma_c \cup \Gamma_N)$, $\Gamma_O = \partial\Omega \cap \{x = 1\} \subset \Gamma_N$. Compare the meshes generated by an energy-norm error estimator for the primal solution, and by the weighted error estimator. Also compare their quality for approximating the misfit $J(u)$ and the control $q$.

# Chapter 9

# Time-Dependent Problems

In this chapter, we will apply the general theory developed in the preceding sections to evolution problems of the following general form: Find $u \in V$ satisfying

$$(\partial_t u, \psi) + a(u)(\psi) = f \quad \forall \psi \in W, \quad t \in (0, T],$$
$$u(0) = u^0. \tag{9.1}$$

Here, $a(\cdot)(\cdot)$ is a semilinear form representing the spatial part of the operator and $V$ and $W$ are certain space-time Hilbert spaces for solutions and test functions, respectively. The discretization will be in the framework of a global *space-time* Galerkin method in the form of a dG-FEM or cG-FEM (*discontinuous* or *continuous* Galerkin finite element method) as already discussed in Chapter 2. Formulation (9.1) represents problems of both parabolic as well as hyperbolic type. The material of this chapter is mainly taken from Hartmann [69], Bangerth [14], and Bangerth and Rannacher [16].

## 9.1 Galerkin discretization

For the discretization, we split the time interval $I := (0, T]$ into half-open subintervals $I_n = (t_{n-1}, t_n]$ of length $k_n := t_n - t_{n-1}$,

$$0 = t_0 < \dots < t_n < \dots < t_N = T.$$

At each time level $t_n$, let $V_h^n$ be appropriate finite dimensional spaces defined on spatial meshes $\mathbb{T}_h^n$ which may vary from time step to time step. Extending the spatial meshes $\mathbb{T}_h^n$ constant in time to the time slabs $\bar{\Omega} \times I_n$, we obtain meshes $\mathbb{T}_h^k$ of the space-time domain $Q_T = \Omega \times I$, which consist of $d+1$-dimensional cubes $Q_K^n := K \times I_n$ (see Figure 9.1). The change of the spatial mesh from one time slab to the next one is accomplished by allowing hanging nodes which due to the possible discontinuity of trial and test functions in time may carry their own unknowns.

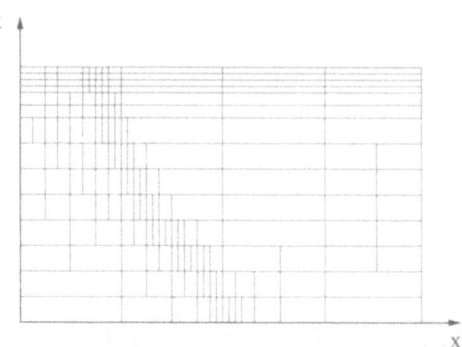

Figure 9.1: *A space-time mesh with hanging nodes in time.*

Then, for fixed integers $r, s \geq 0$, we define the trial spaces $V_h^k$ and test spaces $W_h^k$ by

$$V_h^k := \big\{ \varphi : \bar{Q}_T \to \mathbb{R}, \ \varphi(\cdot, t)_{|\bar{\Omega}} \in V_h^n, \ t \in I_n, \ \varphi(\cdot, 0)_{\bar{\Omega}} \in V_h^0,$$
$$\varphi(x, \cdot)_{|I_n} \in P_r, \ x \in \bar{\Omega} \big\},$$
$$W_h^k := \big\{ \psi : \bar{Q}_T \to \mathbb{R}, \ \psi(\cdot, t)_{|\bar{\Omega}} \in V_h^n, \ t \in I_n, \ \psi(\cdot, 0)_{\bar{\Omega}} \in V_h^0,$$
$$\psi(x, \cdot)_{|I_n} \in P_s, \ x \in \bar{\Omega} \big\}.$$

For functions from these spaces and their continuous analogues, we define

$$v^{n+} = \lim_{t \downarrow t_n} v(t), \quad v^{n-} = \lim_{t \uparrow t_n} v(t), \quad [v]^n = v^{n+} - v^{n-}.$$

At this point the trial functions in $V_h^k$ and the test functions in $W_h^k$ may be both discontinuous in time. Later on, we will consider a situation in which time-continuity of trial functions is required.

With the above notation, we introduce the space-time semilinear form

$$A(u)(\psi) = \sum_{n=1}^N \int_{I_n} \big\{ (\partial_t u, \psi) + a(u)(\psi) - (f, \psi) \big\} \, dt$$
$$+ \sum_{n=2}^N ([u]^{n-1}, \psi^{n-1+}) + (u^{0+}, \psi^{0+}).$$

Then, any sufficiently smooth solution $u$ of the *continuous* problem satisfies the variational relation

$$A(u)(\psi) = (u^0, \psi^{0+}) \quad \forall \psi \in W. \tag{9.2}$$

Notice that here, the initial condition $u_{|t=0} = u^0$ is imposed in variational form. Then, the *space-time* Galerkin method constructs $u_h \in V_h^k$, by requiring that

$$A(u_h)(\psi_h) = (u_h^0, \varphi_h^{0+}) \quad \forall \psi_h \in W_h^k. \tag{9.3}$$

By construction, the error $e := u - u_h$ satisfies Galerkin orthogonality, i.e.

$$A(u)(\psi_h) - A(u_h)(\psi_h) = 0, \quad \psi_h \in W_h^k. \tag{9.4}$$

Suppose that the error is to be estimated in terms of some output functional $J(\cdot)$. To this end, we again employ a dual problem *backward in time* just as in Chapter 2 for ODEs. With our general notation the corresponding dual solution $z$ is determined by the linearized problem

$$A'(u)(\varphi, z) = J'(u)(\varphi) \quad \forall \varphi \in V. \tag{9.5}$$

Then, by the general theory of Chapter 6, we have the following simplified error representation:

$$J(u) - J(u_h) = -A(u_h)(z - \psi_h) + \mathcal{R}_h^{(2)}, \tag{9.6}$$

for arbitrary $\psi_h \in W_h^k$, with the second-order remainder term

$$\mathcal{R}_h^{(2)} = \int_0^1 \left\{ A''(u_h + se)(e, e, z) - J''(u_h + se)(e, z) \right\} s\, ds.$$

From this abstract result, we can derive concrete error estimates. Here, we concentrate on the following special cases applied to two different *linear* model problems:

- The parabolic heat equation, discretized by the dG(0)-FEM where $r = s = 0$ with trial and test functions *discontinuous* in time, which is related to the *backward Euler scheme*.

- The hyperbolic wave equation, discretized by the cG(1)-FEM where $r = 1$ and $s = 0$ with trial functions *continuous* but test functions *discontinuous* in time, which is related to the *implicit midpoint rule*.

In view of the above discussion, the a posteriori error analysis for the space-time Galerkin method for nonstationary problems seems very much to parallel that for the space Galerkin method for stationary problems. So why do we have to go through this again? The reason is the inherent *anisotropy* of the space-time meshes since a constant time step is used within each time slab $\bar{\Omega} \times I_n$. In fact, the present situation may be viewed as a model case for the organization of *anisotropic* mesh refinement where here the anisotropy is in space versus time. Separating the effects of the different scales in space and time requires some care in deriving the a posteriori error estimates; see the discussion in Section 4.4.

## 9.2 A parabolic model problem: the heat equation

As a first example, we consider the heat conduction problem

$$\begin{aligned} \partial_t u - \Delta u &= f \quad \text{in } Q_T := \Omega \times I, \\ u_{|t=0} &= u^0 \quad \text{on } \Omega, \quad u_{|\partial\Omega} = 0 \quad \text{on } I, \end{aligned} \tag{9.7}$$

where $\Omega \subset \mathbb{R}^2$ and $I = (0, T]$. This model is used to describe diffusive transport of energy or certain species concentrations. In concrete applications the heat-conductivity may vary in space, i.e., the diffusion operator has the form $\nabla \cdot \{a\nabla\}$, and in front of the time derivative there may be a density coefficient $\rho \neq$ const.

We continue using the notation of space-time meshes introduced above. In the following, we particularly consider an approximation using shape and test functions which are both piecewise bilinear *continuous* in space and piecewise constant *discontinuous* in time, the so-called dG(0)-FEM. Accordingly, we define the finite element spaces

$$V_h^k := \big\{ \varphi : \bar{Q}_T \to \mathbb{R}, \ \varphi(\cdot, t)_{|\bar{\Omega}} \in V_h^n, \ t \in I_n, \ \varphi(\cdot, 0)_{\bar{\Omega}} \in V_h^0,$$
$$\varphi(x, \cdot)_{|I_n} \in P_0, \ x \in \bar{\Omega} \big\},$$

and set $W_h^k := V_h^k$. Due to the time-discontinuity of the test functions, the variational formulation of the dG(0)-FEM reduces to a sequence of time steps, for $n = 1, 2, ..., N$,

$$\int_{I_n} \big\{ (\nabla u_h, \nabla \psi_h) - (f, \psi_h) \big\} \, dt + (u_h^{(n-1)+} - u_h^{(n-1)-}, \psi_h) = 0 \quad \forall \psi_h \in V_h^n,$$

or, setting $u_h^n := u_h^{n-} \in V_h^n$,

$$(u_h^n, \psi_h) + k_n (\nabla u_h^n, \nabla \psi_h) = (u_h^{n-1}, \psi_h) + \int_{I_n} (f, \psi_h) \, dt \quad \forall \psi_h \in V_h^n,$$

which resembles the *backward Euler scheme* with integral evaluation of the right-hand side $f$.

In order to illustrate the meaning of the abstract result (9.6) for the present situation, we consider control of the error with respect to the following two different output functionals, one being *global* and the other one *local* in time:

1. Space-time $L^2$-norm error $J(e) = \|e\|_{Q_T}$ with the corresponding functional

$$J(\varphi) := \|e\|_{Q_T}^{-1} \int_I (\varphi, e)_\Omega \, dt.$$

The associated dual problem running backward in time reads as follows:

$$-\partial_t z - \Delta z = \|e\|_{Q_T}^{-1} e \quad \text{in } \Omega \times I,$$
$$z_{|t=T} = 0 \ \text{in } \Omega, \ z_{|\partial\Omega} = 0 \ \text{on } I, \tag{9.8}$$

or written in variational form,

$$A'(u)(\varphi, z) = (\varphi, e)_{Q_T} \|e\|_{Q_T}^{-1} \quad \forall \varphi \in W.$$

2. Spatial $L^2$-norm error $J(e) = \|e^{N-}\|_\Omega$, at the end time $T = t_N$, with the corresponding output functional

$$J(\varphi) := \|e^{N-}\|_\Omega^{-1}(\varphi^{N-}, e^{N-}).$$

The associated dual problem is

$$-\partial_t z - \Delta z = 0 \quad \text{in } \Omega \times I,$$
$$z_{|t=T} = \|e^{N-}\|_\Omega^{-1} e^{N-} \quad \text{in } \Omega, \quad z_{|\partial\Omega} = 0 \quad \text{on } I, \tag{9.9}$$

or in variational form

$$A'(u)(\varphi, z) = (\varphi^{N-}, e^{N-})_\Omega \|e^{N-}\|_\Omega^{-1} \qquad \forall \varphi \in W.$$

Note that for the present equation the form $A(\cdot)(\cdot)$ is affine-linear and therefore the derivative form $A'(u)(\cdot, \cdot)$ is actually independent of $u$. For both cases, the general result (9.6) yields the a posteriori error representation

$$J(e) = \sum_{n=1}^{N} \sum_{K \in \mathbb{T}_h^n} \left\{ (R_h^k, z - I_h^k z)_{Q_K^n} + (r_h^k, z - I_h^k z)_{\partial K \times I_n} \right. \tag{9.10}$$
$$\left. - ([u_h]^{n-1}, (z - I_h^k z)^{(n-1)+})_K \right\},$$

with an appropriate approximation $I_h^k z \in V_h^k$ and the local residuals

$$R_{h|K}^k := f - \partial_t u_h + \Delta u_h, \qquad r_{h|\Gamma}^k := \begin{cases} \frac{1}{2}[\partial_n u_h], & \text{if } \Gamma \subset \partial T \backslash \partial\Omega, \\ 0, & \text{if } \Gamma \subset \partial\Omega. \end{cases}$$

Here, we use the natural interpolation $I_h^k z \in V_h^k$ which is defined by

$$\int_{I_n} I_h^k z(a, t) \, dt = \bar{z}(a), \qquad \bar{z}(x) := \int_{I_n} z(x, t) \, dt, \quad x \in \bar{\Omega},$$

for all nodal points $a$ of the mesh $\mathbb{T}_h^n$. Observing that the time-integrated equation residual $\bar{R}_h^k = \bar{f} - \partial_t U + \Delta U$ as well as the jump-residual $r_h^k$ are constant in time, the a posteriori error representation can be rewritten in the form

$$J(e) = \sum_{n=1}^{N} \sum_{K \in \mathbb{T}_h^n} \left\{ (f - \bar{f}, z - I_h^k z)_{Q_K^n} + (\bar{R}_h^k, \bar{z} - I_h^k z)_{Q_K^n} \right. \tag{9.11}$$
$$\left. + (r_h^k, \bar{z} - I_h^k z)_{\partial K \times I_n} - ([u_h]^{n-1}, (z - I_h^k z)^{(n-1)+})_K \right\}.$$

Notice that replacing the dual solution $z$ on the right-hand side by its time-average $\bar{z}$ will be essential in separating the temporal and spatial error indicators in the error estimates to be derived below. From the above error representation, we conclude the following result:

**Proposition 9.1.** *For the approximation of the heat conduction problem by the dG(0)-FEM, there holds the a posteriori error estimate*

$$|J(e)| \leq \eta_\omega := \sum_{n=1}^{N} \sum_{K \in \mathbb{T}_h^n} \left\{ \rho_{K,h}^n \, \omega_{K,h}^n + \rho_{K,k}^n \, \omega_{K,k}^n \right\}, \tag{9.12}$$

*where the cell-wise residuals and weights are defined by*

1. *spatial terms:*

$$\rho_{K,h}^n := \left( \|\bar{R}_h^k\|_{Q_K^n}^2 + h_K^{-1} \|r_h^k\|_{\partial K \times I_n}^2 \right)^{1/2},$$

$$\omega_{K,h}^n := \left( \|\bar{z} - I_h^k z\|_{Q_K^n}^2 + h_K \|\bar{z} - I_h^k z\|_{\partial K \times I_n}^2 \right)^{1/2},$$

2. *temporal terms:*

$$\rho_{K,k}^n := \left( \|f - \bar{f}\|_{Q_K^n}^2 + k_n^{-1} \|[u_h]^{n-1}\|_K^2 \right)^{1/2},$$

$$\omega_{K,k}^n := \left( \|z - I_h^k z\|_{Q_K^n} + k_n \|(z - I_h^k z)^{(n-1)+}\|_K^2 \right)^{1/2}.$$

*Remark* 9.2. We note that in the error estimate (9.12) the effect of the space discretization is separated from that of the time discretization, i.e., on each space-time cell $Q_K^n$ the indicator $\eta_{K,h}^n := \rho_{K,h}^n \omega_{K,h}^n$ can be used to control the spatial mesh width $h_K$, and the indicator $\eta_{K,k}^n := \rho_{K,k}^n \omega_{K,k}^n$ for the time step $k_n$. The details of the resulting adaptation process will be described below.

*Remark* 9.3. The direct use of the above a posteriori error estimate requires the evaluation of the weights $\omega_{K,h}^n$ and $\omega_{K,k}^n$. In principle, this is done analogously to the stationary case by locally post-processing a numerically computed approximation $z_h \in V_h^k$ to the dual solution $z$. Alternatively, one may follow the traditional approach to eliminate $z$ from the error estimate by using appropriate a priori bounds (see Eriksson and Johnson [52, 54, 55, 56], and Eriksson et al. [57]).

In the following, we want to show that from the weighted error estimate (9.12), one obtains 'weight-free' a posteriori error estimates in which the local weights are condensed into just one global stability constant. For this, we have to invest as much a priori knowledge about the dual solution as possible depending on the particular situation. Consequently, the two different cases of target quantities introduced above are considered separately. However, the following two principles will be used throughout.

- Due to the damping properties of the heat equation, local errors in time are expected to decay exponentially which should be reflected in the a posteriori error estimate by a decay factor of the form

$$\sigma(t) := e^{\lambda T} e^{-\lambda t},$$

where $\lambda$ is the smallest eigenvalue of the Laplacian in $V = H_0^1(\Omega)$.

- The use of bilinear elements in space naturally leads to the spatial norm $\|\nabla^2 z\|$ which can be controlled in terms of the $L^2$-initial value $\|z(T)\|$ only by introducing an additional time-weight function

$$\tau(t) := \min\{T-t, \tau_0\},$$

owing to the fact that it takes some time to smooth out irregularities in the initial values. Assuming the relevant time scale to be $T > 1$, we set $\tau_0 = 1$.

At first, we provide the following interpolation estimates.

**Lemma 9.4.** *For the nodal interpolation* $I_h^k z \in V_h^k$, *there holds*

$$\|\bar{z} - I_h^k z\|_{Q_K^n} + h_K^{1/2} \|\bar{z} - I_h^k z\|_{\partial K \times I_n} \leq ch_K^2 \|\nabla^2 z\|_{Q_K^n}, \tag{9.13}$$

$$\|z - I_h^k z\|_{Q_K^n} + k_n^{1/2} \|z - I_h^k z\|_K \leq ch_K^2 \|\nabla^2 z\|_{Q_K^n} + c_I k_n \|\partial_t z\|_{Q_K^n}. \tag{9.14}$$

*Proof.* It suffices to prove the assertion on a reference unit cell $\hat{Q} = \hat{K} \times \hat{I}$ with $h = k = 1$. The interpolation error $w := z - I_h^k z$ has the property

$$\bar{w}(a) := \int_I w(a,t)\, dt = 0,$$

at corner points $a$ of $K$, which implies that

$$\|\bar{w}\|_K + \|\bar{w}\|_{\partial K} \leq c\|\nabla^2 \bar{w}\|_K \leq c\|\nabla^2 w\|_Q.$$

Further, by elementary analysis, there holds

$$w(x,t) = \bar{w}(x) + \int_I \mu(t,s)\partial_t w(x,s)\, ds,$$

$$\mu(t,s) := \left\{ \begin{array}{ll} s, & 0 \leq s < t \\ s-1, & t \leq s \leq 1 \end{array} \right\}.$$

Hence, we have the following first set of inequalities,

$$\|w_{|t=0}\|_K \leq \|\bar{w}\|_K + c\|\partial_t w\|_Q \leq c\{\|\nabla^2 w\|_Q + \|\partial_t w\|_Q\},$$

$$\|w\|_Q \leq \|\bar{w}\|_K + c\|\partial_t w\|_Q \leq c\{\|\nabla^2 w\|_Q + \|\partial_t w\|_Q\}.$$

Next, there holds

$$\|\bar{z} - Iz\|_{\partial K \times I} = \|\bar{w}\|_{\partial K \times I} \leq c\|\nabla^2 w\|_Q.$$

Then, a standard scaling argument from $Q$ to $Q_K^n$ completes the proof. $\qquad\square$

As consequences of Lemma 9.4, we obtain the following estimates for the weights in the error estimate (9.12):

$$\omega_{K,h}^n \leq ch_K^2 \|\nabla^2 z\|_{Q_K^n}, \tag{9.15}$$

$$\omega_{K,k}^n = ch_K^2 \|\nabla^2 z\|_{Q_K^n} + c_I k_n \|\partial_t z\|_{Q_K^n}. \tag{9.16}$$

Using the above estimates, we obtain the following results for the two considered cases:

**Proposition 9.5.** *We have the following a posteriori estimates:*
*1) For the global space-time $L^2$-norm error:*

$$\|e\|_{Q_T} \leq c_I c_S \Big( \sum_{n=1}^{N} \sum_{K \in \mathbb{T}_h^n} \{ h_K^4 (\rho_{K,h}^n)^2 + (h_K^4 + k_n^2)(\rho_{K,k}^n)^2 \} \Big)^{1/2}, \qquad (9.17)$$

*with the stability constant*

$$c_S := c \Big( \int_I \{ \|\nabla^2 z\|^2 + \|\partial_t z\|^2 \} \, dt \Big)^{1/2}.$$

*2) For the end-time $L^2$-norm error:*

$$\|e^{N-}\| \leq c_I c_S \Big( \sum_{n=1}^{N} \sum_{K \in \mathbb{T}_h^n} \sigma_n^{-1} \tau_n^{-1} \times$$

$$\times \{ h_K^4 (\rho_{K,h}^n)^2 + (h_K^4 + k_n^2)(\rho_{K,k}^n)^2 \} \Big)^{1/2}, \qquad (9.18)$$

*with the stability constant*

$$c_S = c \Big( \int_I \sigma(t) \tau(t) \{ \|\nabla^2 z\|^2 + \|\partial_t z\|^2 \} \, dt \Big)^{1/2}.$$

*Proof.* For the space-time $L^2$-norm error, we have

$$\|e\|_{Q_T} \leq \sum_{n=1}^{N} \sum_{K \in \mathbb{T}_h^n} \{ \rho_{K,h}^n \omega_{K,h}^n + \rho_{K,k}^n \omega_{K,k}^n \}$$

$$\leq c_I \sum_{n=1}^{N} \sum_{K \in \mathbb{T}_h^n} \{ \rho_{K,h}^n h_K^2 \|\nabla^2 z\|_{Q_K^n} + + \rho_{K,k}^n (h_K^2 \|\nabla^2 z\|_{Q_K^n} + k_n \|\partial_t z\|_{Q_K^n}) \},$$

which implies the estimate (9.17). The derivation of the estimate (9.18) follows the same line of argument.                                                                  $\square$

In the following lemma, we provide bounds for the stability constants in the estimates (9.17) and (9.18).

**Lemma 9.6.** *The dual solution admits the following a priori bounds:*
*1) For the space-time $L^2$-norm error:*

$$\Big( \int_I \{ \|\partial_t z\|^2 + \|\nabla^2 z\|^2 \} \, dt \Big)^{1/2} \leq c. \qquad (9.19)$$

*2) For the end-time $L^2$-norm error:*

$$\Big( \int_I \sigma(t) \tau(t) \{ \|\partial_t z\|^2 + \|\nabla^2 z\|^2 \} \, dt \Big)^{1/2} \leq c. \qquad (9.20)$$

*The weight functions are* $\tau(t) := \min\{T-t, 1\}$ *and* $\sigma(t) := e^{\lambda T} e^{-\lambda t}$, *where* $\lambda$ *is the smallest eigenvalue of the Laplacian in* $V = H_0^1(\Omega)$. *The constants* $c$ *only depend on the domain* $\Omega$.

*Proof.* For the first case, we multiply the dual equation by $-\partial_t z$ and integrate over $Q_T$, to obtain

$$\int_I \|\partial_t z\|^2 \, dt + \tfrac{1}{2}\|\nabla z(0)\|^2 = \|e\|_{Q_T}^{-1} \int_I (\partial_t z, e)\, dt.$$

Consequently,

$$\|\partial_t z\|_{Q_T}^2 \le \|e\|_{Q_T}^{-1} \|e\|_{Q_T} = 1.$$

Then, using elliptic $H^2$-regularity, $\|\nabla^2 z\| \le c\|\Delta z\|$, the estimate (9.19) follows.

To prove (9.20), notice that $\tau(T) = 0$, $|\tau'| \le 1$, $\sigma(T) = 1$, $\sigma' = -\lambda\sigma$, and, by Poincaré inequality,

$$\lambda\|v\|^2 \le \|\nabla v\|^2, \quad v \in V.$$

First, multiply the dual equation by $\sigma z$, then integrate over $Q_T$,

$$-\tfrac{1}{2}d_t\{\sigma\|z\|^2\} - \tfrac{1}{2}\lambda\sigma\|z\|^2 + \sigma\|\nabla z\|^2 = 0,$$
$$\tfrac{1}{2}\sigma(0)\|z(0)\|^2 + \int_I \sigma\{\|\nabla z\|^2 - \tfrac{1}{2}\lambda\|z\|^2\}\, dt = \tfrac{1}{2}\|z(T)\|^2,$$

and consequently,

$$\int_I \sigma\|\nabla z\|^2 \, dt \le \|z(T)\|^2 = 1.$$

Next, multiply the dual equation by $-\sigma\tau\partial_t z$ and integrate over $I$,

$$\sigma\tau\|\partial_t z\|^2 - \tfrac{1}{2}d_t\{\sigma\tau\|\nabla z\|^2\} + \tfrac{1}{2}\sigma'\tau\|\nabla z\|^2 + \tfrac{1}{2}\sigma\tau'\|\nabla z\|^2 = 0,$$
$$\int_I \sigma\tau\|\partial_t z\|^2 \, dt + \tfrac{1}{2}\sigma(0)\tau(0)\|\nabla z(0)\|^2 + \tfrac{1}{2}\int_I \sigma\{\tau' - \lambda\tau\}\|\nabla z\|^2 \, dt = 0.$$

Combine the last estimates and use again elliptic regularity,

$$\int_I \sigma\tau\{\|\partial_t z\|^2 + \|\nabla^2 z\|^2 \, dt \le c \int_I \sigma\|\nabla z\|^2\}\, dt \le c.$$

The proof is complete. □

## Numerical results

The results of space-time mesh adaptation by the a posteriori error estimator for the dG(0)-FEM is illustrated by a test with a known exact solution representing a smooth rotating bump on the unit square $\Omega = (0,1) \times (0,1)$, corresponding to a heat source moving around a disk. For the mesh adaptation, we start from the a posteriori error estimate

$$|J(e)| \leq \eta_\omega := \sum_{n=1}^{N} \sum_{K \in \mathbb{T}_h^n} \left\{ \rho_{K,h}^n \, \omega_{K,h}^n + \rho_{K,k}^n \, \omega_{K,k}^n \right\}, \qquad (9.21)$$

and introduce the local error indicators

$$\eta_{K,h}^n := \rho_{K,h}^n \, \omega_{K,h}^n, \quad \eta_{K,k}^n := \rho_{K,k}^n \, \omega_{K,k}^n, \quad \eta_k^n := \sum_{K \in \mathbb{T}_h^n} \eta_{K,k}^n.$$

Setting $N$ equal to the number of time steps and $N_n$ equal to the numbers of cells of $\mathbb{T}_h^n$, the spatial mesh size $h_K$ and time step $k_n$ are adapted according to the following balancing criteria:

1. Adaptation in time: Choose time step $k_n$ uniform on mesh $T_h^n$ such that

$$\alpha \frac{\text{TOL}}{2N} \leq \eta_k^n \leq \frac{\text{TOL}}{2N} \qquad (\alpha \approx \tfrac{1}{4}).$$

2. Adaptation in space: Choose $h_K$ such that

$$\beta \frac{k_n}{T} \frac{\text{TOL}}{2N_n} \leq \eta_{K,h}^n \leq \frac{k_n}{T} \frac{\text{TOL}}{2N_n} \qquad (\beta \approx \tfrac{1}{4}).$$

Table 9.1 contains results obtained with the above space-time mesh adaptation for the computation of the end-time $L^2$-norm error. A sequence of corresponding meshes in shown in Figure 9.2. We clearly see the effect of the temporal weights which enforce refinement in space and time only for time levels close to the final time. In the case of the global space-time $L^2$-norm error the adaptation process yields the meshes shown in Figure 9.3. We see that the mesh refinement is slightly stronger at earlier times due to the error propagation in the problem which is captured by the dual weights in the error estimate.

Table 9.1: *Simultaneous adaptation of spatial and temporal mesh size (N number of time-steps, $N_{\max}$ maximum number of mesh-cells; from Hartmann [69].*

| $N$ | $N_{\max}$ | $J(e)$ | $\eta_\omega$ | $I_{\text{eff}}$ |
|---|---|---|---|---|
| 21 | 256 | $1.20 \cdot 10^{-2}$ | $7.14 \cdot 10^{-2}$ | 5.59 |
| 46 | 760 | $2.91 \cdot 10^{-3}$ | $6.60 \cdot 10^{-3}$ | 2.27 |
| 81 | 3472 | $7.92 \cdot 10^{-4}$ | $1.08 \cdot 10^{-3}$ | 1.36 |
| 119 | 9919 | $3.99 \cdot 10^{-4}$ | $6.64 \cdot 10^{-4}$ | 1.67 |

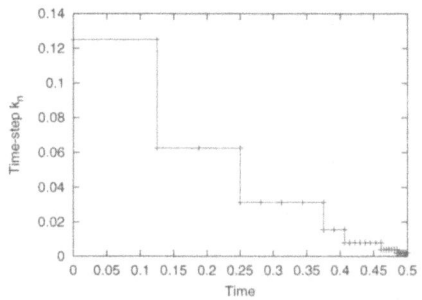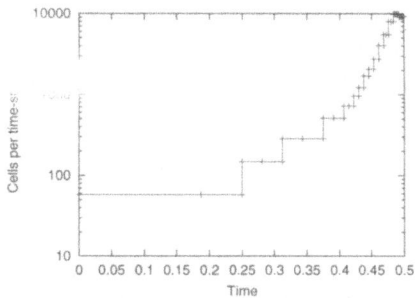

Figure 9.2: *Development of the time-step size (left) and the number $N_n$ of mesh cells (right) over the time interval $I = [0, T = 0.5]$ ; from Hartmann [69].*

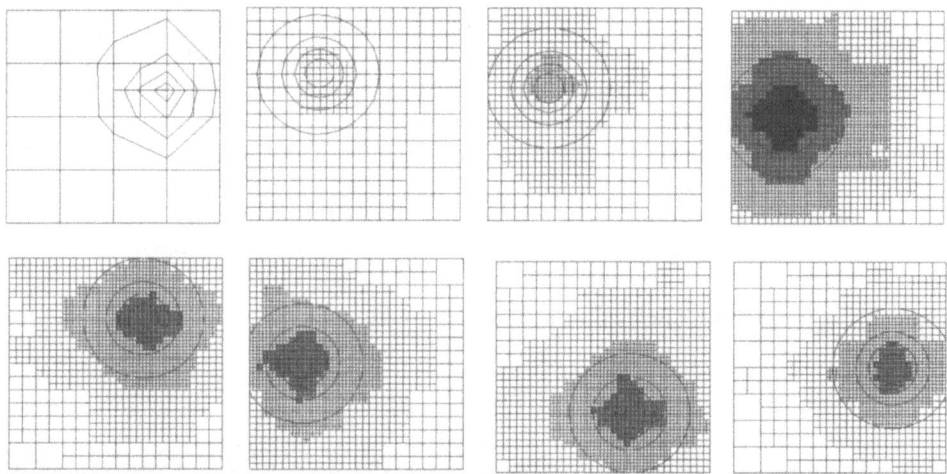

Figure 9.3: *Refined meshes and isolines of the primal solution for the end-time $L^2$-norm error at times $t_n = 0.125, \ldots, 0.5$ (upper row), and for the global $L^2$-norm error at times $t_n = 0.16, \ldots, 1$ (lower row); from Hartmann [69].*

## 9.3  A hyperbolic model problem: the wave equation

As a second example, we consider the acoustic wave equation

$$\partial_t^2 w - \nabla \cdot \{a \nabla w\} = 0 \quad \text{in } Q_T := \Omega \times I,$$
$$w_{|t=0} = w^0, \quad \partial_t w_{|t=0} = v^0 \text{ on } \Omega, \quad w_{|\partial\Omega} = 0 \quad \text{on } I, \tag{9.22}$$

where $\Omega \subset \mathbb{R}^d$ and $I = (0, T]$; the elastic coefficient $a(x)$ may vary in space. This equation occurs in the simulation of acoustic waves in gaseous or fluid media. We consider its solution by a *velocity-displacement* formulation which is obtained by introducing a new velocity variable $v := \partial_t w$. Then, the pair $u = \{w, v\}$ satisfies the system of equations

$$\partial_t w - v = 0,$$
$$\partial_t v - \nabla \cdot \{a \nabla w\} = 0. \tag{9.23}$$

We continue using the notation introduced in the preceding section. On each time slab $Q^n := \Omega \times I_n$, we define intermediate meshes $\bar{\mathbb{T}}_h^n$ which are composed of the mutually finest cells of the neighboring meshes $\mathbb{T}_h^{n-1}$ and $\mathbb{T}_h^n$ defined at discrete times $t_{n-1}, t_n$, respectively, and obtain a decomposition of the time slab into space-time cubes $Q_K^n = K \times I_n, K \in \bar{\mathbb{T}}_h^n$. This construction is used in order to allow continuity in time of the trial functions when the meshes change with time. The discrete trial spaces $V_h^k$ in the space-time domain consist of functions that are $(d+1)$-linear on each space-time cell $Q_K^n$ and *globally continuous* on $Q_T$. This prescription requires the use of 'hanging nodes' if the spatial mesh changes across a time level $t_n$. The corresponding discrete test spaces $W_h^k$ consist of functions that are constant in time on each cell $Q_K^n$, while they are $d$-linear in space and globally continuous on $\Omega$. On these spaces, we introduce the affine-linear form

$$
\begin{aligned}
A(u)(\psi) := \ & (\partial_t w, \psi^v)_{Q_T} - (v, \psi^v)_{Q_T} + (w(0) - w^0, \psi^v(0)) \\
& + (\partial_t v, \psi^w)_{Q_T} + (a \nabla w, \nabla \psi^w)_{Q_T} + (v(0) - v^0, \psi^w(0)),
\end{aligned}
$$

where $\psi = \{\psi^v, \psi^w\}$. The Galerkin method seeks $u_h = \{w_h, v_h\} \in V_h^k$ satisfying

$$A(u_h)(\psi_h) = 0 \quad \forall \psi_h = \{\psi_h^v, \psi_h^w\} \in W_h^k. \tag{9.24}$$

This scheme is a so-called *Petrov-Galerkin method* since trial and test spaces do not coincide. Since the continuous solution $u = \{w, v\}$ also satisfies (9.24), we again have Galerkin orthogonality for the error $e := \{e^w, e^v\}$,

$$A(u)(\psi_h) - A(u_h)(\psi_h) = 0, \quad \psi_h \in V_h^k. \tag{9.25}$$

This time-discretization scheme is called the *cG(1)-FEM* (*continuous* Galerkin method) in contrast to the dG-FEM used in the preceding section. Such a dG-FEM for solving the wave equation is described in Johnson [86]. We note that, from the cG(1) scheme, we can recover the standard Crank-Nicolson scheme in time (combined with a spatial finite element method):

$$
\begin{aligned}
(w^n - w^{n-1}, \psi^v) - \tfrac{1}{2} k_n (v^n + v^{n-1}, \psi^v) &= 0, \\
(v^n - v^{n-1}, \psi^w) + \tfrac{1}{2} k_n (a \nabla(w^n + w^{n-1}), \nabla \psi^w) &= 0.
\end{aligned}
\tag{9.26}
$$

This system can be split into two equations, a discrete Helmholtz equation and a discrete $L^2$ projection.

We will embed the present situation into the above general framework. Let $V$ and $W$ be the natural function Hilbert spaces for the solution and the test functions of the system (9.23). We want to control the error in terms of an output functional of the form

$$J(\varphi) := (j, \varphi^w)_{Q_T}, \quad \varphi = \{\varphi^w, \varphi^v\},$$

with some density function $j(x,t)$ which may be a measure or a Dirac function in space-time. To this end, we again use a duality argument in space-time employing the time-reversed wave equation

$$\partial_t^2 z^w - \nabla\cdot\{a\nabla z^w\} = j \quad \text{in } Q_T,$$
$$z^w|_{t=T} = 0, \quad -\partial_t z^w|_{t=T} = 0 \quad \text{on } \Omega, \qquad (9.27)$$
$$n\cdot a\nabla z^w|_{\partial\Omega} = 0 \quad \text{on } I.$$

Its strong solution $z = \{-\partial_t z^w, z^w\}$ satisfies the variational equation

$$A'(u)(\varphi, z) = J(\varphi) \quad \forall \varphi \in V, \qquad (9.28)$$

where again the derivative form $A'(u)(\cdot, \cdot)$ is actually independent of $u$. From the general theory, we obtain the error representation

$$J(e) = -A'(u)(w_h, z - \psi_h), \qquad (9.29)$$

for arbitrary $\psi_h = \{\psi_h^w, \psi_h^v\} \in W_h^k$. Recalling the definition of the affine-linear form $A(\cdot)(\cdot)$, we obtain

$$|(j, w)_{Q_T}| \le \sum_{n=1}^{N} \sum_{K \in \mathbb{T}_h^n} |(R_h^w, z^v - \varphi_h^v)_{Q_K^n} \atop + (R_h^v, z^w - \varphi_h^w)_{Q_K^n} - (r_h, z^w - \varphi_h^w)_{\partial K \times I_n}|, \qquad (9.30)$$

with the cell residuals

$$R_{h|K}^w := \partial_t w_h - v_h, \qquad R_{h|K}^v := \partial_t v_h - \nabla\cdot\{a\nabla w_h\}$$

and the edge residuals

$$r_{h|\Gamma \times I_n} := \begin{cases} \frac{1}{2} n\cdot[a\nabla w_h], & \text{if } \Gamma \subset \partial K \backslash \partial\Omega, \\ 0, & \text{if } \Gamma \subset \partial\Omega. \end{cases}$$

From this, we infer the following result:

**Proposition 9.7.** *For the cG(1)-FEM applied to the acoustic wave equation, we have the a posteriori error estimate*

$$|J(e)| \leq \eta_\omega := \sum_{n=1}^{N} \sum_{K \in \mathbb{T}_h^n} \{ \rho_K^{n,1} \omega_K^{n,1} + \rho_K^{n,2} \omega_K^{n,2} \}, \tag{9.31}$$

*where the cell-wise residuals and weights are defined by*

$$\rho_K^{n,1} := \left( \|R_h^v\|_{Q_K^n}^2 + h_K^{-1} \|r_h\|_{\partial K \times I_n}^2 \right)^{1/2},$$
$$\omega_K^{n,1} := \left( \|\partial_t z^w - \psi_h^v\|_{Q_K^n}^2 + h_K \|z^w - \psi_h^w\|_{\partial K \times I_n}^2 \right)^{1/2},$$
$$\rho_K^{n,2} := \|R_h^w\|_{Q_K^n},$$
$$\omega_K^{n,2} := \|z^w - \psi_h^w\|_{Q_K^n},$$

*for arbitrary $\{\psi_h^w, \psi_h^v\} \in V_h^k$ .*

Below, we will compare the above weighted error estimator with a simple heuristic energy-norm error estimator measuring the spatial smoothness of $w_h$ :

$$\eta_E := \left( \sum_{n=1}^{N} \sum_{K \in \mathbb{T}_h^n} h_K^2 (\rho_K^{n,2})^2 \right)^{1/2}.$$

### Numerical results

We consider the propagation of an outward traveling wave on $\Omega = (-1,1)^2$ with a strongly heterogeneous coefficient. The boundary and initial conditions are

$$n \cdot \{a \nabla u\} = 0 \quad \text{on } y = 1, \qquad w = 0 \quad \text{on } \partial\Omega \setminus \{y = 1\},$$
$$w_0 = 0, \quad v_0 = \theta(s - r) \exp\left(-|x|^2/s^2\right) \left(1 - |x|^2/s^2\right), \tag{9.32}$$

with $s = 0.02$, and $\theta(\cdot)$ being the jump function (see Figure 9.4).

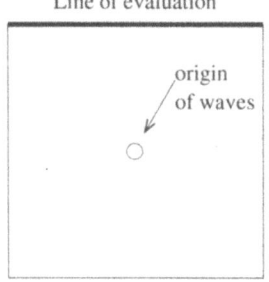
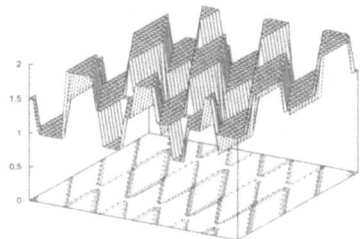

Figure 9.4: *Layout of the domain (left) and structure of the coefficient $a(x)$ (right).*

The lowest frequency in the initial wave field has wavelength $\lambda = 4s$. Hence taking the usual minimum ten grid points per wavelength would yield $62,500$ cells already for the largest wavelength; however, higher frequencies also have a prominent role. If this example is taken as a model of propagation of seismic waves in a faulted region of rock, then the seismograms at the surface, the top line $\Gamma$ of the domain, are to be recorded. A corresponding functional output is

$$ J(w) = \int_0^T \int_\Gamma w(x,t)\, \omega(\xi,t)\, \mathrm{d}\xi\, \mathrm{dt}, $$

with a weight $\omega(\xi,t) = \sin(3\pi\xi)\sin(5\pi t/T)$, and end-time $T = 2$. The frequency of oscillation of this weight is chosen to match the frequencies in the wave field to obtain good resolution of changes.

The results obtained for this output functional using the weighted error estimator $\eta_\omega$ and the energy-norm error estimator $\eta_E$ and corresponding adapted meshes are shown in Table 9.2 and Figure 9.5, respectively. We clearly see the reason for the better efficiency of $\eta_\omega$ which induces mesh refinement only in the domain of dependence of the output functional. In fact, the energy-norm error estimator resolves the wave field well, including reflections from discontinuities in the coefficient, while the weighted error estimator additionally takes into account that the lower parts of the domain lie outside the domain of influence of the target functional if we truncate the time domain at $T = 2$.

Table 9.2: *Results obtained by the DWR method (reference value $J(w) \approx -4.515 \cdot 10^{-6}$), $\Sigma_N = \sum_{n=1}^N N_n$ total number of mesh cells, $N = $ number of time-steps, and $N_n = $ number of spatial mesh cells at $t_n$; from Bangerth [14].*

| weighted estimator | | heuristic estimator | |
|---|---|---|---|
| $\Sigma_N$ | $J(w_h)$ | $\Sigma_N$ | $J(w_h)$ |
| 327,789 | $-2.09 \cdot 10^{-6}$ | 327,789 | $-2.09 \cdot 10^{-6}$ |
| 920,380 | $-4.63 \cdot 10^{-6}$ | 920,380 | $-4.63 \cdot 10^{-6}$ |
| 2,403,759 | $-4.29 \cdot 10^{-6}$ | 2,403,759 | $-4.29 \cdot 10^{-6}$ |
| 1,918,696 | $-4.18 \cdot 10^{-6}$ | 5,640,223 | $-4.39 \cdot 10^{-6}$ |
| 2,975,119 | $-4.438 \cdot 10^{-6}$ | 10,189,837 | $-4.46 \cdot 10^{-6}$ |
| 6,203,497 | $-4.524 \cdot 10^{-6}$ | 17,912,981 | $-4.521 \cdot 10^{-6}$ |
| | | 41,991,779 | $-4.517 \cdot 10^{-6}$ |

*Remark* 9.8. The evaluation of the *a posteriori* error estimate requires a careful approximation of the adjoint solution $z$. Therefore, a higher-order method with biquadratic elements is used for solving the space-time dual problem. A compromise between accuracy in approximating $z$ and work efficiency may be found in computing only the *biquadratic* defect of the *bilinear* approximation and to apply a few steps of defect correction with the low-order system and a 'cheap' higher-order blockwise preconditioning.

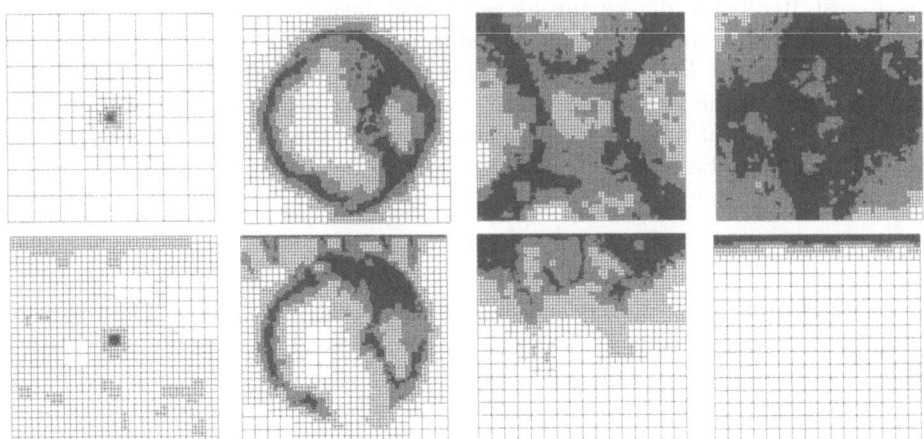

Figure 9.5: *Sequence of refined meshes produced by the energy-norm error estimator (upper row) and by the weighted estimator (lower row), at times $t = 0, \frac{2}{3}, \frac{4}{3}, 2$; from Bangerth [14].*

## 9.4   Exercises

*Exercise* 9.1. The cG(1)-FEM (*continuous Galerkin FEM*) uses piecewise linear continuous-in-time trial functions and discontinuous-in-time piecewise constant test functions. This Petrov-Galerkin method is closely related to the implicit mid-point rule as an ODE solver. Formulate this scheme for discretizing the heat conduction problem and derive an a posteriori estimate for the end-time error $\|e^{N^-}\|$.

*Exercise* 9.2. The wave propagation problem can be written as a first-order system in time also in the form

$$\nabla\cdot\{a\nabla\partial_t w\} - \nabla\cdot\{a\nabla v\} = 0, \quad \partial_t v - \nabla\cdot\{a\nabla w\} = 0.$$

Develop a *weighted* a posteriori error estimate for the discretization of this system by the cG(1)-FEM as described above.

*Exercise* 9.3 *(Practical exercise)*. Repeat the numerical test in Section 9.2 for the heat equation with the right-hand side $f \equiv 1$ and initial value $u^0 = 0$. The goal is to compute the end-time value $u(x,T)$, say for $T = 1$, with best efficiency. What is the relation $N = N(\mathrm{TOL})$ for the number of time steps to be expected on the basis of the weighted error estimator $\eta_\omega$ in (9.21), if the spatial meshes are kept fixed (sufficiently fine), i.e., $\mathbb{T}_h^n = \mathbb{T}_h$, $n \geq 0$? Choosing $T$ larger, this results in the so-called *pseudo-time stepping scheme* for computing an approximation $u_h^\infty \in V_h$ to the solution $u^\infty \in V$ of the stationary limit equation $-\Delta u^\infty = f$.

# Chapter 10

# Applications in Structural Mechanics

In this chapter, we will present some applications of the DWR method to typical problems in structural mechanics. At first, the standard finite element approximation of the linear Lamé-Navier system is considered which is the basic model for small static deformations of elastic bodies. This does not add much to the experience we have already gained before at the Poisson problem. Then, we turn to a more challenging problem, the deformation of an elasto-plastic body assuming linear-elastic and perfectly plastic material. The finite element approximation is rather standard but includes stabilization in order to cope with almost incompressible material behavior. Here, the interesting point is the non-differentiable nonlinearity which occurs if the constrained problem is reformulated as a variational equation. The results presented in this Chapter are taken from Suttmeier [128], and Rannacher and Suttmeier [118, 119, 120, 121].

## 10.1  Approximation of the Lamé-Navier system

The fundamental problem of linear elasticity theory is the Lamé-Navier equation which is the Euler-Lagrange system corresponding to the following minimization problem: Minimize the total energy functional

$$E_{\text{tot}}(u) := \tfrac{1}{2}(C\varepsilon(u), \varepsilon(u)) - (f, u) - (g, u)_{\Gamma_N} \quad \to \quad \min, \tag{10.1}$$

for $u \in V := \{v \in H^1(\Omega)^d, \ v = 0 \text{ on } \Gamma_D\}$, where $\varepsilon(u) = \tfrac{1}{2}(\nabla u + \nabla u^T)$ is the strain tensor corresponding to the displacement vector $u$, and $C$ is a positive definite material tensor. Here and below, we use the standard scalar products of tensor functions, e.g., $(C\varepsilon, \varepsilon) := \sum_{i,j,k,l=1}^{d}(C_{ijkl}\varepsilon_{kl}, \varepsilon_{ij})$.

This model describes the (small) deformation of an elastic body occupying a bounded polygonal or polyhedral domain $\Omega \subset \mathbb{R}^d$ ($d = 2$ or $3$) which is fixed along a part $\Gamma_D$ of its boundary $\partial\Omega$, under the action of a body force $f$, and a surface traction $g$ along $\Gamma_N = \partial\Omega \setminus \Gamma_D$. We assume a linear-elastic and isotropic material law for the stress-strain relation

$$\sigma = C\varepsilon := 2\mu\,\varepsilon^D + \kappa\,\mathrm{tr}(\varepsilon)I,$$

with constants $\mu > 0$ and $\kappa > 0$, where $\varepsilon^D$ denotes the deviatoric part $\varepsilon^D := \varepsilon - \frac{1}{d}\mathrm{tr}(\varepsilon)$ of $\varepsilon$. With the bilinear form (so-called *energy form*)

$$a(\cdot,\cdot) = (C\varepsilon(\cdot),\varepsilon(\cdot)),$$

the *primal variational formulation* of problem (10.1) reads

$$a(u,\psi) = (f,\psi) + (g,\psi)_{\Gamma_N} \qquad \forall\,\psi \in V, \tag{10.2}$$

where the solution space is $V = \{v \in H^1(\Omega)^d,\ v = 0 \text{ on } \Gamma_D\}$. Assuming that $\Gamma_D \neq \emptyset$, by Korn inequality, we have the coercivity estimate

$$a(v,v) =: \|v\|_E^2 \geq \gamma\|v\|_1^2, \quad v \in V,$$

with some constant $\gamma > 0$. In this case problem (10.2) has a unique solution $u$ with bounded *energy norm* $\|u\|_E$.

We consider the discretization of (10.2) by $d$-linear finite elements as defined in Section 3.1, with subspaces $V_h \subset V$. The meshes $\mathbb{T}_h$ are assumed to match the decomposition $\partial\Omega = \Gamma_D \cup \Gamma_{\mathbb{N}}$. The Galerkin approximation reads

$$a(u_h,\psi_h) = (f,\psi_h) + (g,\psi_h)_{\Gamma_N} \qquad \forall\,\psi_h \in V_h. \tag{10.3}$$

As before, the error $e := u - u_h$ satisfies the Galerkin orthogonality relation

$$a(e,\psi_h) = 0, \quad \psi_h \in V_h,$$

from which we obtain the usual a priori energy-norm error estimate

$$\|e\|_E \leq \inf_{\psi \in V_h} \|u - \psi_h\|_E.$$

Once the approximate displacement $u_h$ has been computed the corresponding discrete stress is obtained by $\sigma_h := C\varepsilon(u_h)$.

Next, we turn to the a posteriori error analysis. For a given (linear) output functional $J(\cdot)$ defined on $V$, we introduce the dual problem

$$a(\varphi,z) = J(\varphi) \qquad \forall\,\varphi \in V. \tag{10.4}$$

Now, we proceed analogously as before in the scalar case. Taking $\varphi = e$ in (10.4) and using Galerkin orthogonality, we have

$$J(e) = a(e,z) = a(e, z - \psi_h), \quad \psi_h \in V_h.$$

Splitting the global integration over $\Omega$ into the contributions of the cells $K \in \mathbb{T}_h$ and integrating by parts cell-wise yields again that

$$J(e) = \sum_{K \in \mathbb{T}_h} \left\{ (-\nabla \cdot C\varepsilon(e), z - \psi_h)_K + (n \cdot C\varepsilon(e), z - \psi_h)_{\partial K} \right\}.$$

Observing $-\nabla \cdot C\varepsilon(u) = f$ and the continuity of $n \cdot C\varepsilon(u)$ across inter-element edges, we obtain the error representation

$$J(e) = \sum_{K \in \mathbb{T}_h} \left\{ (R_h, z - \psi_h)_K + (r_h, z - \psi_h)_{\partial K} \right\}, \tag{10.5}$$

with the cell residuals $R_{h|K} := f + \nabla \cdot C\varepsilon(u_h)$ and the edge residuals

$$r_{h|\Gamma} := \begin{cases} \frac{1}{2} n \cdot [C\varepsilon(u_h)], & \text{if } \Gamma \subset \partial K \setminus \partial\Omega, \\ 0, & \text{if } \Gamma \subset \Gamma_D, \\ g - n \cdot C\varepsilon(u_h), & \text{if } \Gamma \subset \Gamma_N. \end{cases}$$

From the error identity (10.5), we can infer the following a posteriori error estimate:

**Proposition 10.1.** *For the finite element approximation of the Lamé-Navier equations, we have the a posteriori error estimate*

$$|J(e)| \leq \eta_\omega := \sum_{K \in \mathbb{T}_h} \rho_K \omega_K, \tag{10.6}$$

*with the cell residuals $\rho_K$ and weights $\omega_K$ defined by*

$$\rho_K := \left( \|R_h\|_K^2 + \tfrac{1}{2} h_K^{-1} \|r_h\|_{\partial K}^2 \right)^{1/2},$$
$$\omega_K := \left( \|z - \psi_h\|_K^2 + \tfrac{1}{2} h_K \|z - \psi_h\|_{\partial K}^2 \right)^{1/2},$$

*for arbitrary $\psi_h \in V_h$.*

We note that in the estimate (10.6) the cell and edge residual terms are separated, by which we may lose possible error cancellation between the cells. However, this is not harmful for the cases considered in this chapter as neither primal nor dual solutions will be oscillatory.

Analogously as in the scalar case in Section 3.2, we can derive from (10.6) the usual *energy-norm error estimate*

$$\|e\|_E \leq \eta_E := c_S c_I \left( \sum_{K \in \mathbb{T}_h} \rho_K^2 \right)^{1/2}, \tag{10.7}$$

where the stability constant $c_S$ is proportional to the maximum condition number of the material tensor $C$ on $\bar{\Omega}$.

## Numerical results

In order to test the a posteriori error estimator derived above for the Lamé-Navier equations, we consider a square elastic disc with a crack (see Figure 10.1) which is subjected to a constant boundary traction acting along half of the upper boundary. Along the right-hand and lower parts of the boundary the disc is clamped and along the remaining part of the boundary (including the crack) it is left free. The solution of this problem has a stress singularity of the type $\sigma \approx r^{-1/2}$ (expressed in polar coordinates $(r, \theta)$ at the tip of the crack). The material parameters are chosen as commonly used for aluminum, i.e., $\mu = 80193.80 \, N/mm^2$ and $\kappa = 164206 \, N/mm^2$. The surface traction is of size $g = 0.1 \, N/m^2$.

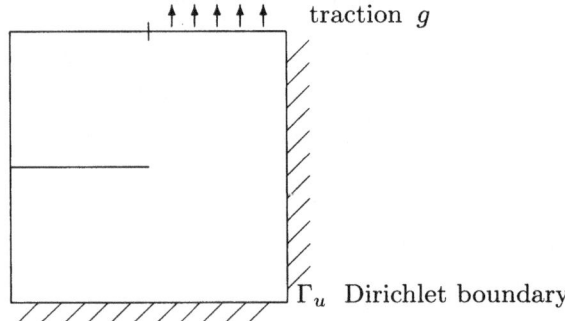

Figure 10.1: *Configuration of the crack-domain model.*

The goal of the computation is the mean normal stress over the Dirichlet part $\Gamma_D$ of the boundary, i.e.,

$$J(u) = \int_{\Gamma_u} n \cdot \sigma(u) \cdot n \, ds \,.$$

Again, this singular functional has to be regularized with regularization parameter chosen as $\varepsilon = TOL$, see the approach described in Section 3.3. Alternatively, one could transform it by integration by parts into a volume-oriented form

$$J(u) = \int_{\Omega} \{\nabla \cdot \sigma \cdot \bar{n} + \sigma : \nabla \bar{n}\} \, dx = \int_{\Omega} \{-f \cdot \bar{n} + \sigma : \nabla \bar{n}\} \, dx,$$

where $\sigma : \nabla \bar{n}$ denotes the natural scalar product of matrices, and $\bar{n}$ is a Lipschitz continuous extension of the normal unit vector on $\Gamma_D$ into $\Omega$ which vanishes on $\partial\Omega \setminus \Gamma_D$. The advantage of this reformulation is that now $J(\cdot)$ is well defined on the whole solution space $V$, which makes its evaluation on $V_h$ more robust and accurate. An analogous procedure will be used in Chapter 11 for evaluating the drag and lift coefficients of a body in a viscous flow. A trivial special case of this had already been seen in Example 3.5.

We use the following notation for the *relative error* $E_{\text{rel}}$ and the *effectivity index* $I_{\text{eff}}$, where $\sigma_{\text{ref}}$ is a reference stress obtained on a very fine mesh:

$$E_{\text{rel}} := \left| \frac{J_\varepsilon(\sigma_h - \sigma_{\text{ref}})}{J_\varepsilon(\sigma_{\text{ref}})} \right|, \quad I_{\text{eff}} := \left| \frac{\eta_\omega}{J_\varepsilon(\sigma_h - \sigma_{\text{ref}})} \right|.$$

The results of the computations for this situation are shown in Table 10.1 and corresponding optimized meshes in Figure 10.2. We see that $\eta_E$ induces more mesh refinement in those areas where the solution has its irregularities, i.e. around the tip of the slit and at the boundary where the boundary condition jumps. In contrast to that, $\eta_\omega$ puts less emphasis onto these regions but additionally induces mesh refinement along $\Gamma_D$ where particular accuracy is needed for evaluating the target quantity.

Table 10.1: *Results for $\eta_E$ (left) and $\eta_\omega$ (right); from Rannacher and Suttmeier [118].*

| L | N | $J(u_h)$ | $E_{\text{rel}}$ |
|---|------|---------|---------|
| 3 | 1180 | 0.01936 | 0.0188 |
| 4 | 2659 | 0.02053 | 0.0139 |
| 5 | 6193 | 0.02154 | 0.0096 |
| 6 | 13423 | 0.02232 | 0.0064 |
| 7 | 31336 | 0.02281 | 0.0043 |
| 8 | 65332 | 0.02315 | 0.0029 |

| L | N | $J(u_h)$ | $E_{\text{rel}}$ | $I_{\text{eff}}$ |
|---|------|---------|---------|------|
| 3 | 1060 | 0.02114 | 0.0113 | 1.95 |
| 4 | 2113 | 0.02216 | 0.0070 | 1.96 |
| 5 | 4435 | 0.02280 | 0.0044 | 1.92 |
| 6 | 8830 | 0.02320 | 0.0027 | 1.86 |
| 7 | 15886 | 0.02343 | 0.0017 | 1.79 |
| 8 | 29947 | 0.02359 | 0.0010 | 1.79 |

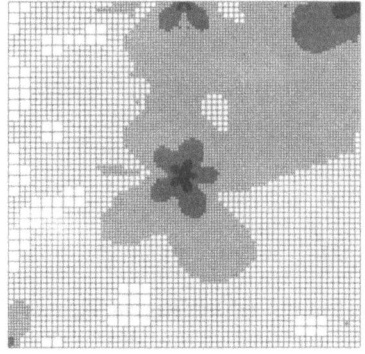 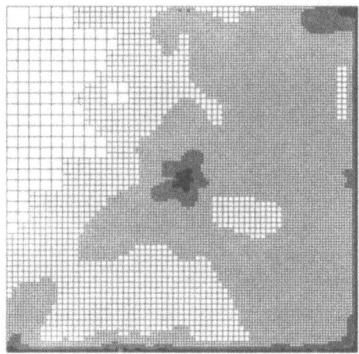

Figure 10.2: *Meshes adapted by $\eta_E$ (left) and $\eta_\omega$ (right); from Rannacher and Suttmeier [118].*

## 10.2   A model problem in elasto-plasticity theory

Let us now turn to plastic materials. The fundamental problem in the static deformation theory of linear-elastic perfect-plastic material (so-called *Hencky model*)
reads in variational form as follows: Minimize the total energy functional

$$E_{\text{tot}}(u) := \tfrac{1}{2}(C\varepsilon(u), \varepsilon(u)) - (f, u) - (g, u)_{\Gamma_N} \quad \to \quad \min, \qquad (10.8)$$

for $u \in V := \{v \in H^1(\Omega)^d, \ v = 0 \text{ on } \Gamma_D\}$ under the pointwise constraint that

$$|\sigma^D| \leq \sigma_0 \quad \text{in } \Omega,$$

where $\sigma^D := \sigma - \frac{1}{d}\text{tr}(\sigma)I$ is again the deviatoric part of the stress $\sigma = C\varepsilon$, and
$\sigma_0 > 0$ is a material-dependent bound. This model describes the deformation of
an elasto-plastic body occupying a bounded domain $\Omega \subset \mathbb{R}^d$ ($d=2$ or 3) which is
fixed along a part $\Gamma_D$ of its boundary $\partial\Omega$, under the action of a body force $f$
and a surface traction $g$ along $\Gamma_N = \partial\Omega\backslash\Gamma_D$. For the elastic part of the body, we
assume again a linear-elastic and isotropic material law, $C\varepsilon := 2\mu\,\varepsilon^D + \kappa\,\text{tr}(\varepsilon)I$,
with constants $\mu > 0$ and $\kappa > 0$.

  The natural variational formulation of the elasto-plasticity problem is that
of a variational inequality. Since our general approach for deriving a posteriori
error estimates as developed in Chapter 6 requires the underlying problem to be a
variational *equation*, we prefer to rewrite the system (10.8) as a possibly nonlinear
variational equation. This is actually possible for the elasto-plasticity problem due
to the particular structure of the inequality constraint. To this end, we introduce
the nonlinear operator

$$C(\varepsilon(u)) = \Pi(2\mu\varepsilon^D(u)) + \kappa\,\text{tr}(\varepsilon(u))I,$$

where $\Pi$ is the tensor-valued projector into the ball of radius $\sigma_0$, which is the
maximum strain the material can support:

$$\Pi(2\mu\varepsilon^D(u)) = \begin{cases} 2\mu\varepsilon^D(u), & \text{if } |2\mu\varepsilon^D(u)| \leq \sigma_0, \\ \dfrac{\sigma_0}{|\varepsilon^D(u)|}\varepsilon^D(u), & \text{if } |2\mu\varepsilon^D(u)| > \sigma_0. \end{cases}$$

Then, the primal variational formulation of (10.8) has the form

$$a(u)(\psi) := (C(\varepsilon(u)), \varepsilon(\psi)) - (f, \psi) - (g, \psi)_{\Gamma_N} = 0 \qquad \forall\psi \in V. \qquad (10.9)$$

The nonlinearity in $C(\cdot)$ is only Lipschitz continuous which makes this problem
difficult. Nevertheless, we will see that the DWR method also works in this situation, although lacking a rigorous theoretical justification. Note that the semilinear
form $a(\cdot)(\cdot)$ is monotone.

  The finite element approximation of problem (10.9) reads

$$a(u_h)(\psi_h) = 0 \qquad \forall\psi_h \in V_h. \qquad (10.10)$$

This nonlinear algebraic problem is solved by a damped Newton iteration. With a discrete displacement $u_h$, we then associate a corresponding stress $\sigma_h$ by

$$\sigma_h = \Pi(2\mu\,\varepsilon^D(u_h)) + \kappa\,\mathrm{tr}(\varepsilon(u_h))I.$$

We have to be careful in applying the DWR method to this situation. Since the semilinear form $a(\cdot)(\cdot)$ is not differentiable, the formal approach developed in Chapter 6 to define a dual problem and compute an error representation does not apply directly. However, we can use it as a guideline for deriving, at least heuristically, weighted a posteriori error estimators for the approximating scheme (10.10). Then, numerical practice has to determine whether this works. Some theoretical justification can be found in the observation that the non-differentiability of the function $C(\cdot)$ occurs only in the elastic-plastic transition zone which can be expected to be lower-dimensional. This may explain the apparent success of the method seen below.

Let $J(\cdot)$ be a (linear) output functional with respect to which the discretization error $e = u - u_h$ is to be controled. Then, the formal application of the approach of Chapter 6 leads us to an error representation of the form

$$J(e) = -a(u_h)(z - \psi_h) + \mathcal{R}_h^{(2)}, \quad \psi_h \in V_h, \tag{10.11}$$

where the remainder term $\mathcal{R}_h^{(2)}$ is quadratic in the error $e$ in regions where the function $C(\cdot)$ is twice differentiable. The problematic area is the elastic-plastic transition zone where regularity breaks down. The residual term in the error identity has the form

$$-a(u_h)(z - \psi_h) = \sum_{K \in \mathbb{T}_h} \left\{ (R_h, z - \psi_h)_K + (r_h, z - \psi_h)_{\partial K} \right\},$$

with the cell and edge residuals defined as usual by

$$R_{h|K} = f + \nabla \cdot C(\varepsilon(u_h)),$$

$$r_{h|\Gamma} = \begin{cases} \frac{1}{2} n \cdot [C(\varepsilon(u_h))], & \text{if } \Gamma \subset \partial K \setminus \partial\Omega, \\ g - n \cdot C(\varepsilon(u_h)), & \text{if } \Gamma \subset \Gamma_N, \\ 0, & \text{if } \Gamma \subset \partial\Omega \setminus \Gamma_N. \end{cases}$$

Clearly, the dual solution $z$ cannot be defined in the usual way by a dual problem involving first-order derivatives of the form $a(\cdot)(\cdot)$. Instead, we adopt a heuristic approach and define the following approximate dual problem

$$(\tilde{C}'(\varepsilon(u_h))\varepsilon(\varphi), \varepsilon(\tilde{z})) = J(\varphi) \quad \forall \varphi \in V. \tag{10.12}$$

Here, the derivative $\tilde{C}'(\cdot)$ is defined piecewise as follows:

$$\tilde{C}'(\tau)\varepsilon := \begin{cases} C\varepsilon, & \text{if } |\tau^D| \le \sigma_0, \\ \dfrac{\sigma_0}{|\tau^D|}\left\{ \varepsilon^D - \dfrac{((\tau^D)^T : \varepsilon^D)\tau^D}{|\tau^D|^2} \right\} + \kappa\,\mathrm{tr}(\varepsilon)I, & \text{if } |\tau^D| > \sigma_0. \end{cases}$$

With this notation we have the following (heuristic) result:

**Proposition 10.2.** *For the finite element approximation of the Hencky model, we have the (heuristic) a posteriori error estimate*

$$|J(e)| \approx \eta_\omega = \sum_{K \in \mathbb{T}_h} \rho_K \omega_K, \tag{10.13}$$

*with the cell residuals $\rho_K$ and weights $\omega_K$ defined by*

$$\rho_K := \{\|R_h\|_K^2 + \tfrac{1}{2}h_K^{-1}\|r_h\|_{\partial K}^2\}^{1/2},$$
$$\omega_K := \{\|\tilde{z} - \psi_h\|_K^2 + \tfrac{1}{2}h_K\|\tilde{z} - \psi_h\|_{\partial K}^2\}^{1/2},$$

*for arbitrary $\psi_h \in V_h$.*

The a posteriori error indicator $\eta_\omega$ may be evaluated by either one of the strategies described previously. For simplicity, the linearization of the adjoint problem uses a decomposition of the mesh domain $\Omega_h = \cup\{K \in \mathbb{T}_h\}$ into its 'elastic' and 'plastic' regions $\Omega_h^e$ and $\Omega_h^p$ defined by

$$\Omega_h^e := \cup\{K \in \mathbb{T}_h, \, |2\mu\varepsilon^D|_K| \le \sigma_0\}, \qquad \Omega_h^p := \Omega \setminus \Omega_h^e.$$

Then, in the dual problem $\tilde{C}'$ is approximated piecewise constant with respect to this decomposition taking one or the other branch of the definition of $\tilde{C}'$ above. Using this notation, the adjoint solution $\tilde{z}$ is approximated by the solution $\tilde{z}_h \in V_h$ of the discretized dual problem

$$(\varepsilon(\varphi_h), \tilde{C}_h'(\varepsilon(u_h))^*\varepsilon(\tilde{z}_h)) = J(\varphi_h) \quad \forall\varphi_h \in V_h, \tag{10.14}$$

The evaluation of the coefficient $\tilde{C}_h'(\varepsilon(u_h))^*$ on cells in the elastic-plastic transition zone is usually done by simple numerical integration. This may appear to be a rather crude approximation, but it works in practice. The reason may be that the critical situation only occurs in cells intersecting the elastic-plastic transition zone, which is apparently lower-dimensional. The weights $\omega_K$ are then again approximated as described in Section 4.1 yielding an approximate error estimator $\tilde{\eta}_\omega(u_h)$. We emphasize that the computation of the dual solution requires us to solve only a *linear* problem and normally amounts only to a small fraction of the total cost within the Newton iteration for the nonlinear primal problem.

The error indicator $\tilde{\eta}_\omega$ has been used for automatic mesh adaptation in the test calculations presented below. Its performance will be compared against that of the following two other more traditional heuristic error estimators:

**The ZZ-error indicator (see Zienkiewicz and Zhu [137])**
An approximation $\sigma \approx \mathcal{M}_h\sigma_h$ to $\sigma$ is constructed by local nodal-value averaging which yields the following a posteriori error estimator:

$$\|e^\sigma\| \approx \eta_{ZZ} = \left( \sum_{K \in \mathbb{T}_h} \|\mathcal{M}_h\sigma_h - \sigma_h\|_K^2 \right)^{1/2}.$$

The nodal value determining $\mathcal{M}_h \sigma_h$ at a point of the triangulation is obtained by averaging the cell-wise average values of $\sigma_h$ of those cells having this point in common (see Figure 10.3).

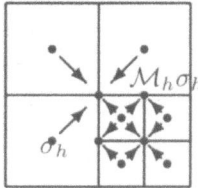

Figure 10.3: *Nodal-value averaging in the ZZ-error estimator.*

**An energy-norm error estimator (see Johnson and Hansbo [88, 89])**
This heuristic energy-norm error estimator is based on decomposing the domain $\Omega$ into *discrete* plastic and elastic zones, $\Omega = \Omega_h^p \cup \Omega_h^e$, and approximating

$$\|e^\sigma\| \approx \eta_E = c_I \Big( \sum_{K \in \mathbb{T}_h} \eta_K^2 \Big)^{1/2},$$

with the local error indicators

$$\eta_K^2 := \begin{cases} h_K^2 \rho_K^2, & \text{if } K \subset \Omega_h^e, \\ \rho_K \|\mathcal{M}_h \sigma_h - \sigma_h\|_K, & \text{if } K \subset \Omega_h^p, \end{cases}$$

where $\rho_K$ are the same residual terms as used in (10.13).

## Numerical results

**Test case 1.** For our first numerical test, we again choose the model configuration 'square plate with slit' shown in Figure 10.4, where now in the elasto-plastic case $\sigma_0 = \sqrt{2/3} \cdot 450$. Under the prescribed surface traction a plastic region develops. Let the goal again be the accurate computation of the mean normal stress over the Dirichlet part $\Gamma_D$ of the boundary, i.e.,

$$J(u) = \int_{\Gamma_u} n \cdot \sigma(u) \cdot n \, ds,$$

respectively its regularized form.

The corresponding results on the basis of the weighted error estimator $\eta_\omega$ are shown in Table 10.2. This refinement indicator proves efficient even on relatively coarse meshes which indicates that the strategy of evaluating the weights $\omega_K$ computationally works also for the present irregular nonlinear problem. The comparison with the other two error estimators $\eta_{ZZ}$ and $\eta_E$ is made in Figure 10.5 which supports the expected superiority of the weighted error estimators.

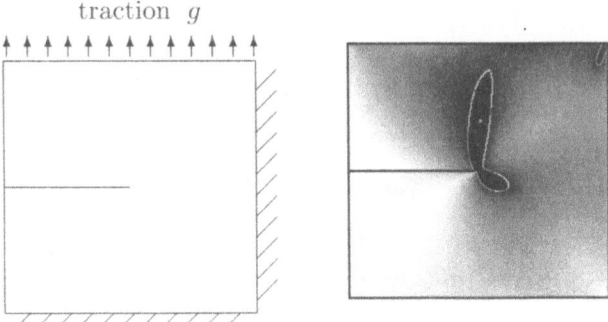

Figure 10.4: *Geometry of the test problem 'square plate with slit' and plot of $|\sigma^D|$ (plastic regions black, elastic-plastic transition zone white); from Rannacher and Suttmeier [119].*

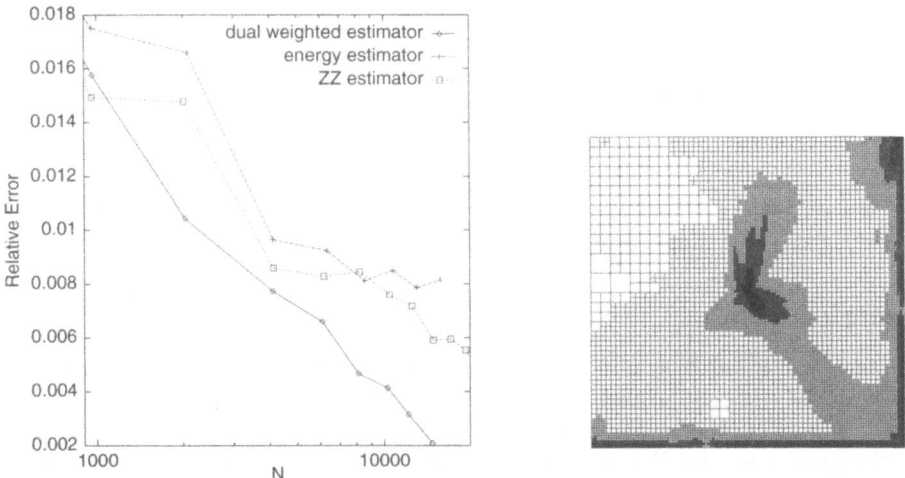

Figure 10.5: *Mesh efficiencies of the different estimators (left) and structure of optimized grid with $N \approx 8,100$ (right); from Rannacher and Suttmeier [119].*

**Test case 2.** The second test uses the benchmark 'square plate with a hole', the configuration of which is shown in Figure 10.6. In this geometrically two-dimensional model again perfectly plastic material behavior is assumed with plane-strain approximation, i.e., $\varepsilon_{i3} = 0$. Due to symmetry the computation is restricted to a

Table 10.2: *Results obtained by the weighted error estimator $\eta_\omega$ ; from Rannacher and Suttmeier [119].*

| N | $J(\sigma_h)$ | $|J(e)/J(\sigma)|$ | $I_{\text{eff}}$ |
|---|---|---|---|
| 1000 | 222.2 | $1.6 \cdot 10^{-2}$ | 1.7 |
| 2000 | 223.4 | $1.0 \cdot 10^{-2}$ | 2.1 |
| 4000 | 224.0 | $7.7 \cdot 10^{-3}$ | 1.7 |
| 8000 | 224.7 | $4.7 \cdot 10^{-3}$ | 1.6 |
| 16000 | 225.3 | $2.1 \cdot 10^{-3}$ | 1.8 |
| $\infty$ | 225.8 | | |

quarter-domain. The material parameters are again chosen as above in the purely elastic case, and $\sigma_0 = \sqrt{2/3} \cdot 450$. The boundary traction is given in the form $g(t) = tg_0$, $g_0 = 100$, $t \in [0,6]$. For the stationary Hencky model, the calculations are performed with one load step from $t = 0$ to $t = 4.5$; for details on this benchmark see the survey article Rannacher and Suttmeier [121]. Among the quantities to be computed are the point-values of the displacements $u_1$ and $u_2$ at the points $P_1$ and $P_5$, respectively.

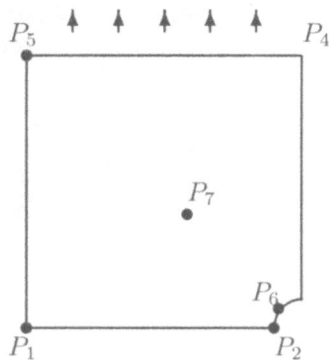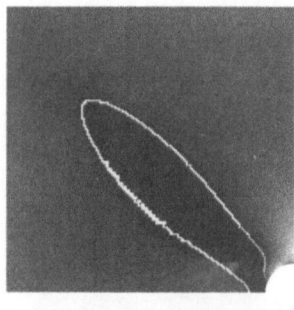

Figure 10.6: *Geometry of the benchmark problem and plot of $|\sigma^D|$ (elastic-plastic transition zone white) computed on a mesh with $N \approx 10,000$ cells; from Rannacher and Suttmeier [121].*

The solutions on very fine (adapted) meshes with about $200,000$ cells are taken as reference solutions $u_{\text{ref}}$ for determining the relative errors $E_{\text{rel}}$ and the effectivity indices $I_{\text{eff}}$ of the error estimator. The results shown in Figure 10.7 indicate a very good efficiency of the constructed meshes. Figure 10.8 shows plots of the weights corresponding to these point-error evaluations. One sees a quite different behavior with respect to the elastic-plastic transition zone which is responsible for the different meshes produced for the two cases.

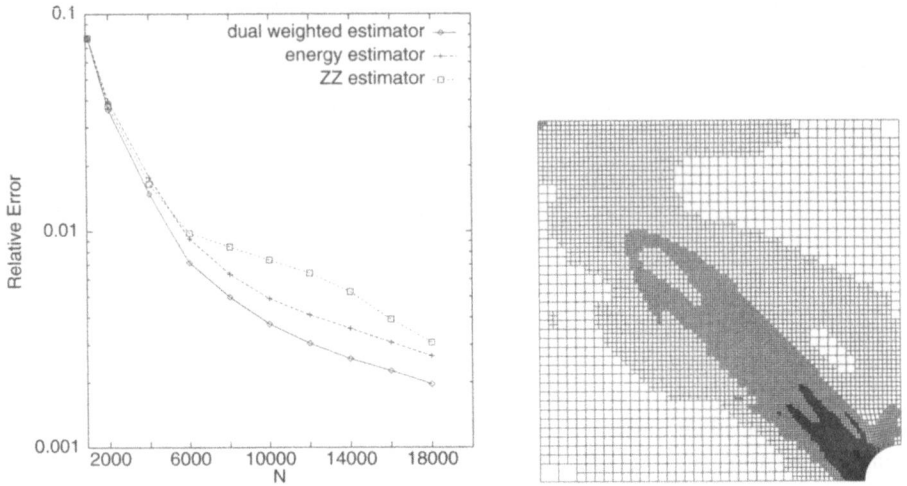

Figure 10.7: *Error for computing $u_1(P_5)$ using the different error estimators and an 'optimized' mesh with about 10,000 cells; from Rannacher and Suttmeier [121].*

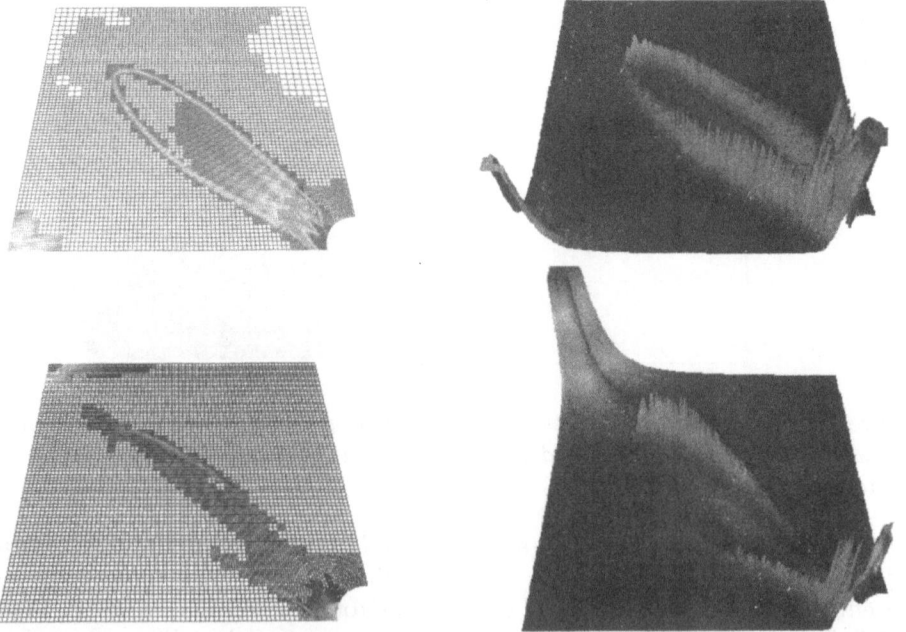

Figure 10.8: *Meshes for computing $u_1(P_1)$ (top) and $u_2(P_5)$ (bottom), and corresponding distributions of weights $\omega_K$; from Rannacher and Suttmeier [121].*

*Remark* 10.3. In the plastic region the material behavior is almost incompressible, which causes stability problems for the discretization based on the formulation (10.9). In order to cope with this problem, a stabilized finite element discretization may be employed by introducing an auxiliary pressure variable (see Suttmeier [129]). We briefly describe this displacement-pressure formulation. The finite element subspace $V_h \subset V$ is supplemented by a subspace $Q_h \subset Q = L^2(\Omega)$ for the discrete pressure. Then, the corresponding discrete problem seeks $\{u_h, p_h\} \in V_h \times Q_h$ such that

$$
\begin{aligned}
a^{\mathrm{dev}}(u_h)(\psi_h) - (p_h, \nabla \cdot \psi_h) &= 0, \\
(\nabla \cdot u_h, \chi_h) + (\kappa^{-1} p_h, \chi_h) &= 0,
\end{aligned}
\tag{10.15}
$$

for all $\{\psi_h, \chi_h\} \in V_h \times Q_h$, where

$$
a^{\mathrm{dev}}(u_h)(\psi_h) := (\Pi(2\mu\varepsilon^D(u_h)), \varepsilon(\psi_h)) - (f, \psi_h) - (g, \psi_h)_{\Gamma_N}
$$

is the deviatoric part of the semilinear form $a(\cdot)(\cdot)$. Here, we choose the 'pressure space' $Q_h$ of equal-order as the 'displacement space' $V_h$, i.e., it consists also of continuous, piecewise (isoparametric) bilinear functions. This discretization would be unstable due to a stiff displacement-pressure coupling for $\kappa \gg 1$, i.e., it does not satisfy the discrete analogue of the so-called *inf-sup stability condition*

$$
\inf_{q \in L} \sup_{v \in H} \frac{(q, \nabla \cdot v)}{\|q\| \, \|\varepsilon(v)\|} \geq \beta,
\tag{10.16}
$$

with a reasonable constant $\beta > 0$ uniformly in $\kappa$, and as $h \to 0$. In order to avoid this difficulty, we introduce a *pressure stabilization* leading to the scheme

$$
\begin{aligned}
a^{\mathrm{dev}}(u_h)(\psi_h) - (p_h, \nabla \cdot \psi_h) &= 0, \\
(\nabla \cdot u_h, q_h) + (\kappa^{-1} p_h, q_h) + \sum_{K \in \mathbb{T}_h} \delta_K (\nabla p_h, \nabla q_h)_K &= 0,
\end{aligned}
\tag{10.17}
$$

for all $\{\psi_h, q_h\} \in V_h \times Q_h$. The stabilization parameter is chosen as $\delta_K = \alpha h_K^2$, with some fixed constant $\alpha \sim 1$. For the so modified scheme, there holds

$$
\sup_{u_h \in V_h} \frac{(p_h, \nabla \cdot u_h)}{\|\varepsilon(u_h)\|} + \left( \sum_{K \in \mathbb{T}_h} \delta_K \|\nabla p_h\|_K^2 \right)^{1/2} \geq \gamma \|p_h\|, \quad p_h \in Q_h,
\tag{10.18}
$$

which together with the monotonicity of the nonlinear form $a(\cdot)(\cdot)$ guarantees the discrete stability. It is demonstrated in Suttmeier [129] that the approximation based on this displacement-pressure formulation yields improved accuracy compared to the pure displacement formulation.

*Remark* 10.4. The adaptive finite element method described above for solving the *stationary* Hencky problem in perfect plasticity has been extended to the *quasi-stationary* Prandtl-Reuss model in Rannacher and Suttmeier [120, 121]. There, the time stepping is done by the backward Euler scheme and it is shown that the resulting incremental errors in each time step can accumulate at most linearly. The adaptive control of the time step through a space-time duality argument has been considered by Hansbo [67]. The incorporation of the elasto-plasticity problem into the general framework of the DWR method relies on its reformulation as a non-linear variational *equation*. For the application of duality techniques for a posteriori error estimation in the direct approximation of variational *inequalities*, we refer to Johnson [85] and particularly to the recent papers of Blum and Suttmeier [35, 36], where an extension of the DWR method is used.

## 10.3   Exercises

*Exercise* 10.1. Formulate one step of the quasi-Newton iteration for solving the nonlinear equation

$$a(u_h)(\psi_h) := (C(\varepsilon(u_h)), \varepsilon(\psi_h)) - (f, \psi_h) - (g, \psi_h)_{\Gamma_N} = 0 \qquad \forall \psi_h \in V_h,$$

in the finite element space $V_h$ and show the solvability of this linear system.

*Exercise* 10.2. Prove the *inf-sup stability estimate* (10.18),

$$\sup_{u_h \in V_h} \frac{(p_h, \nabla \cdot u_h)}{\|\varepsilon(u_h)\|} + \Big( \sum_{K \in \mathbb{T}_h} \delta_K \|\nabla p_h\|_K^2 \Big)^{1/2} \geq \gamma \|p_h\|, \quad p_h \in Q_h,$$

for the pressure-stabilized scheme

$$a^{\mathrm{dev}}(u_h)(\psi_h) - (p_h, \nabla \cdot \psi_h) = 0,$$

$$(\nabla \cdot u_h, q_h) + \kappa^{-1}(p_h, q_h) + \sum_{K \in \mathbb{T}_h} \delta_K (\nabla p_h, \nabla q_h)_K = 0.$$

Use the 'continuous' inf-sup-stability estimate

$$\sup_{u \in V} \frac{(p, \nabla \cdot u)}{\|\varepsilon(u)\|} \geq \gamma \|p\|, \quad p \in Q.$$

as a starting point.

# Chapter 11

# Applications in Fluid Mechanics

In this chapter, we apply the DWR method to problems in fluid mechanics which are all related to the 'incompressible' Navier-Stokes equation for pairs $u = \{v, p\}$:

$$\mathcal{A}(u) := \begin{bmatrix} -\nu \Delta v + v \cdot \nabla v + \nabla p - f \\ \nabla \cdot v \end{bmatrix} = 0.$$

This equation models viscous, incompressible fluid flow described by the velocity $v$ and pressure $p$. We will consider the following proto-typical problems:

- The *computation* of a functional value $J(u)$, e.g. the drag coefficient, from the solution $u$ of

$$\mathcal{A}(u) = 0. \tag{11.1}$$

- The *optimization* of $J(u)$, e.g. the minimization of the drag coefficient, by varying a control parameter $q$ under the constraint

$$\mathcal{A}(u) + \mathcal{B}q = 0, \tag{11.2}$$

  with an operator $\mathcal{B}$ that couples the control variable to the state variable.

- The determination of the stability of a stationary 'base solution' $\hat{u}$ by solving the *stability eigenvalue problem*

$$\mathcal{A}'(\hat{u})u - \lambda \mathcal{M}u = 0. \tag{11.3}$$

We will see that all these situations can be treated using the general framework developed in Chapters 6–8. The results presented in this chapter are taken from Becker [20, 21], and Heuveline and Rannacher [78]; see also the survey article Becker and Rannacher [31].

## 11.1   Computation of drag and lift in a viscous flow

We consider the stationary Navier-Stokes system

$$\mathcal{A}(u) = 0, \tag{11.4}$$

for $u := \{v, p\}$, where $v$ is the velocity vector and $p$ the scalar pressure, describing the motion of a viscous, isothermal, incompressible fluid. The viscosity parameter $\nu > 0$ and the volume force $f$ are given, and the density is normalized to one. As a prototypical situation we have in mind a configuration such as shown in Figure 11.1. In such a case the system (11.4) is supplemented by the boundary conditions

$$v|_{\Gamma_{\text{rigid}}} = 0, \quad v|_{\Gamma_{\text{in}}} = v^{\text{in}}, \quad \nu\partial_n v - np|_{\Gamma_{\text{out}}} = 0, \tag{11.5}$$

which model the usual 'no-slip' condition along the rigid walls $\Gamma_{\text{rigid}}$, a prescribed inflow along $\Gamma_{\text{in}}$ with maximum value $\bar{U}$, and a 'free' outflow along $\Gamma_{\text{out}}$. These are standard boundary conditions for pipe flow which have proven successful in many numerical simulations (for a discussion of the Neumann-type outflow boundary condition see Heywood et al. [75]). The viscosity is chosen such that the Reynolds number is small enough, e.g. $\text{Re} = \bar{U}^2 \text{D}/\nu = 20$, to guarantee stationarity of the flow.

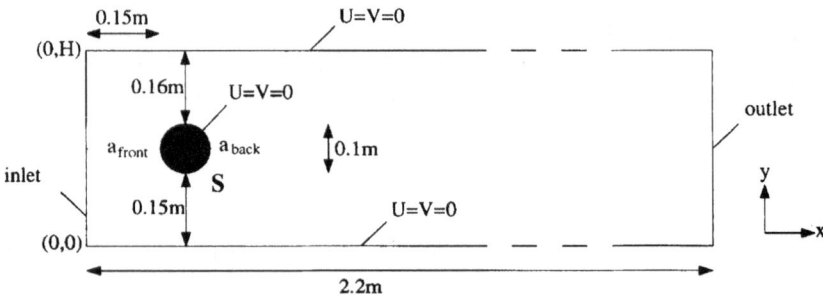

Figure 11.1: *Configuration of the cylinder-flow model problem.*

Suppose that the goal is to compute with high accuracy the *drag and lift coefficients* of some body immersed in the flow, e.g. the circle in our 2-d model configuration. These scalar quantities are defined by

$$c_{\text{drag}} := \frac{2}{\bar{U}^2 D} \int_S n^T (2\nu\tau - pI) e^{(1)} \, ds, \qquad c_{\text{lift}} := \frac{2}{\bar{U}^2 D} \int_S n^T (2\nu\tau - pI) e^{(2)} \, ds,$$

where $S$ is the surface of the cylinder, $D$ its diameter, $\bar{U}$ the reference velocity, and $\tau = \frac{1}{2}(\nabla v + \nabla v^T)$ the strain tensor, and $e^{(1)}$ and $e^{(2)}$ are the unit vectors in downwind and crosswind directions of the flow.

*Remark* 11.1. The computation of the drag and lift coefficients from the above formulas is numerically not very stable since first derivatives of the solution have to be evaluated along the boundary. Hence, it is preferable to rewrite these formulas in a volume-oriented form by integration by parts (see also the exercise at the end of this chapter):

$$c_{\text{drag}} = \frac{2}{\bar{U}^2 D} \int_{\partial\Omega} (v \cdot \nabla v - f) \cdot \bar{e}^{(1)} + (2\nu\tau - pI) : \nabla \bar{e}^{(1)} \, dx, \qquad (11.6)$$

and analogously for $c_{\text{lift}}$, where $\bar{e}^{(1)}$ is an extension of the directional vector $e^{(1)}$ from $S$ to $\Omega$ with support along $S$. On the continuous level these representations are identical but they differ for the discrete solution $u_h$. Practical experience has shown that using the representation (11.6) can gain a whole order of accuracy compared to the surface-oriented form (see Giles et al. [66] and Becker [20]).

### Finite element discretization

For discretizing the Navier-Stokes problem, we use a standard finite element method based on the same trial spaces as already used for the elliptic model problems before. The natural variational formulation of this problem uses the product function space $V := H \times L$ for the velocity $v$ and pressure $p$, where

$$L := L^2(\Omega), \quad H := \{v \in H^1(\Omega)^d : v_{|\Gamma_{\text{in}} \cup \Gamma_{\text{rigid}}} = 0\}.$$

For arguments $u = \{v, p\}$, $\psi = \{\psi^v, \psi^p\} \in V$, we introduce the semilinear form

$$a(u)(\psi) := \nu(\nabla u, \nabla \psi^v) + (v \cdot \nabla v, \psi^v) - (p, \nabla \cdot \psi^v) - (f, \psi^v) - (\nabla \cdot v, \psi^p).$$

With this notation the Navier-Stokes problem (11.4)–(11.5) can be written in compact form as follows: Seek $u \in u^{\text{in}} + V$, satisfying

$$a(u)(\psi) = 0 \qquad \forall \psi \in V, \qquad (11.7)$$

where $u^{\text{in}}$ is a suitable extension of the inflow data $\{v^{\text{in}}, 0\}$ to the whole flow domain $\Omega$. Due to the 'free outflow' boundary condition, the existence of a solution to this problem for our Reynolds number does not follow from the standard theory (developed for Dirichlet boundary conditions), but is clearly suggested by practical experience (see Heywood and Rannacher [75], and also Rannacher [116] for a discussion of this issue).

The variational form (11.7) of the Navier-Stokes problem is the basis of the finite element Galerkin discretization. There are various types of finite element schemes available in the literature. Here, we consider only a particularly simple one, the so-called $Q_1/Q_1$ *Stokes element* which uses continuous piecewise bilinear shape functions on quadrilateral meshes $\mathbb{T}_h$ for both quantities, velocity and pressure as indicated in Figure 11.2. The corresponding finite element subspaces are denoted by

$$L_h \subset L, \quad H_h \subset H, \quad V_h := H_h \times L_h.$$

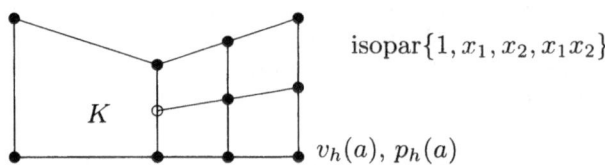

Figure 11.2: *Nodal points of the $Q_1/Q_1$ Stokes element.*

Due to the 'equal-order interpolation' used in this Stokes element it lacks stability for $h \to 0$, i.e., the crucial *inf-sup stability condition*

$$\inf_{q_h \in L_h} \sup_{v_h \in H_h} \frac{(q_h, \nabla \cdot v_h)}{\|q_h\| \, \|\nabla v_h\|} =: \beta_h \geq \beta > 0,$$

is not satisfied for any $\beta$ independent of $h$. The same situation has already occurred in Chapter 10 when dealing with incompressible material behavior. This problem can be cured by adding a pressure stabilization. This is accomplished here together with the also necessary transport stabilization by a standard 'least-squares' approach; see Hughes et al. [82], Hughes and Brooks [80], and Johnson [83]. To this end, we define the stabilization operator

$$\mathcal{S}(u)\psi := \begin{bmatrix} v \cdot \nabla \psi^v + \nabla \psi^p \\ \nabla \cdot \psi^v \end{bmatrix},$$

and the mesh- and parameter-dependent bilinear form

$$(\varphi, \psi)_\delta := \sum_{K \in \mathbb{T}_h} \delta_K \, (\varphi, \psi)_K.$$

With this notation, we introduce the stabilized Galerkin approximation of the Navier-Stokes problem: Find $u_h \in u_h^{\text{in}} + V_h$ satisfying

$$a_\delta(u_h)(\psi_h) := a(u_h)(\psi_h) + (\mathcal{A}(u_h), \mathcal{S}(u_h)\psi_h)_\delta = 0 \qquad \forall \psi_h \in \mathcal{V}_h. \qquad (11.8)$$

This stabilization is fully consistent, i.e., the exact solution $u$ satisfies

$$a_\delta(u)(\psi_h) = 0, \quad \psi_h \in V_h.$$

It provides the following features:

- stabilization of pressure: $\qquad\qquad\qquad \delta_K(\nabla p_h, \nabla \psi_h^p)_K,$
- stabilization of transport: $\qquad\qquad\quad \delta_K(v_h \cdot \nabla v_h, v_h \cdot \nabla \psi_h^v)_K,$
- stabilization of mass conservation: $\quad \delta_K(\nabla \cdot u_h, \nabla \cdot \psi_h^v)_K.$

The stabilization parameters $\delta_K$ are chosen adaptively according to

$$\delta_K = \alpha \big( \nu h_K^{-2} + \beta |v_h|_{\infty;K} h_K^{-1} \big)^{-1},$$

with hand-tuned parameters $\alpha = \frac{1}{12}$ and $\beta = \frac{1}{6}$. The discretized Navier-Stokes system (11.8) is solved by a quasi-Newton iteration as described in Section 6.2.

## A posteriori error estimation

From the general theory of Chapter 6, we obtain the following a posteriori error representation for the present situation:

$$J(u) - J(u_h) \approx \eta_\omega := \tfrac{1}{2}\rho(u_h)(z - z_h) + \tfrac{1}{2}\rho^*(z_h)(u - u_h). \qquad (11.9)$$

For computing the functional value $J(u)$ the dual solution $z = \{z^v, z^p\} \in V$ is determined by the continuous dual problem

$$a'(u)(\varphi, z) = J(\varphi) \quad \forall \varphi \in V, \qquad (11.10)$$

with the bilinear form

$$\begin{aligned} a'(u)(\varphi, z) := {}& \nu(\nabla\varphi^v, \nabla z^v) + (\varphi^v \cdot \nabla v, z^v) + (v \cdot \nabla\varphi^v, z^v) \\ & - (\nabla\cdot\varphi^v, z^p) - (\varphi^p, \nabla\cdot z^v). \end{aligned}$$

This problem reads in strong form like

$$\begin{bmatrix} -\nu\Delta z^v - v\cdot\nabla z^v - (\nabla\cdot v)z^v + (\nabla v)z^v + \nabla z^p \\ \nabla\cdot z^v \end{bmatrix} = \begin{bmatrix} j^v \\ j^p \end{bmatrix}, \qquad (11.11)$$

with an $L^2$ representation $\{j^v, j^p\}$ of the functional $J(\cdot)$. For the technical details of the derivation of this dual problem and also its variant associated to the modified drag functional, we refer to Becker [20]. The solvability of the dual problem can be argued on the basis of assumptions about the local uniqueness of the primal solution $u$. Since this is not really relevant for practical computations, we will not pursue this question. The discrete dual solution $z_h \in V_h$ is obtained from a stabilized version of problem (11.10). The precise form of this discrete dual problem cannot be discussed here in detail since it very much depends on the particular finite element scheme used and on the structure of the problem, i.e. geometry, Reynolds number, etc. In the present situation, transposition, i.e. formation of the dual problem, and stabilization do not commute. We have chosen the option of first to transpose and then to add stabilization in the transposed system in order to ensure its stable solution. The alternative strategy of transposing the whole stabilized system requires differentiation of the stabilization terms and may lead to numerically unstable dual problems (see Becker [20, 22]).

*Remark* 11.2. Note that in the strong form (11.11) of the dual problem, we have kept the term $(\nabla\cdot v)z^v$, though it vanishes for the solenoidal primal velocity $v$. However, the discrete dual problem is derived from the variational form (11.10), with $v$ replaced by its discrete analogue $v_h$ which is usually not exactly divergence-free. Therefore, the corresponding equation residual of $z_h$ *does* contain the term $(\nabla\cdot v_h)z_h^v$. Hence, in this case the discrete dual problem we have to work with slightly differs from the approximation obtained from the discretization of the continuous dual problem in its strong form (see also Remark 6.13). This has to be observed in determining the cell and edge residuals corresponding to $z_h$.

In the present case the primal residual has the following form:

$$\rho(u_h)(z - \psi_h) := \sum_{K \in \mathbb{T}_h} \left\{ (R_h, z^v - \psi_h^v)_K + (r_h, z^v - \psi_h^v)_{\partial K} \right.$$

$$\left. + (\nabla \cdot v_h, z^p - \psi_h^p)_K + \dots \right\},$$

with the cell and edge residuals defined by

$$R_{h|K} := f + \nu \Delta v_h - v_h \cdot \nabla v_h - \nabla p_h,$$

$$r_{h|\Gamma} := \begin{cases} \frac{1}{2}[\nu \partial_n v_h - n p_h], & \text{if } \Gamma \not\subset \partial \Omega, \\ 0, & \text{if } \Gamma \subset \Gamma_{\text{rigid}} \cup \Gamma_{\text{in}}, \\ -\nu \partial_n v_h + n p_h, & \text{if } \Gamma \subset \Gamma_{\text{out}}. \end{cases}$$

Here, the dots stand for additional terms representing the errors caused by the polygonal approximation of the curved boundary, the discrete approximation of the inflow, and the stabilization. All these error components are suppressed in our error estimator since on properly refined meshes they have turned out to be negligibly small. Note that in the present case, we use continuous pressure approximations and hence the jump terms $[n p_h]$ vanish. For the sake of clarity, we keep them in the definition of the edge residuals.

Analogously, we obtain for the dual residual:

$$\rho^*(u_h, z_h)(u - \varphi_h) := \sum_{K \in \mathbb{T}_h} \left\{ (v - \varphi_h^v, R_h^*)_K + (v - \varphi_h^v, r_h^*)_{\partial K} \right.$$

$$\left. + (p - \varphi_h^p, \nabla \cdot z_h^v)_K + \dots \right\},$$

with the cell and edge residuals defined by

$$R_{h|K}^* = j + \nu \Delta z_h^v + (\nabla \cdot v_h) z_h^v + v_h \cdot \nabla z_h^v - (\nabla v_h)^T z_h^v - \nabla z_h^p,$$

$$r_{h|\Gamma}^* = \begin{cases} \frac{1}{2}[\nu \partial_n z_h^v + (v_h \cdot n) z_h^v - n z_h^p] & \text{if } \Gamma \not\subset \partial \Omega, \\ -\nu \partial_n z_h^v - (v_h \cdot n) z_h^v + n z_h^p & \text{if } \Gamma \subset \Gamma_{\text{out}}, \\ 0 & \text{if } \Gamma \subset \Gamma_{\text{rigid}} \cup \Gamma_{\text{in}}. \end{cases}$$

The practical evaluation of the error estimator

$$|\eta_\omega| = \left| \tfrac{1}{2}\rho(u_h)(z - z_h) + \tfrac{1}{2}\rho^*(z_h)(u - u_h) \right|$$

requires us to provide approximations to the primal and dual solutions $u$ and $z$, respectively, which are not from the discrete space $V_h$. As before, we post-process the computed approximations $u_h, z_h \in V_h$ by patch-wise biquadratic interpolation obtaining

$$(z - z_h)_{|K} \approx (I_{2h}^{(2)} z_h - z_h)_{|K}, \qquad (u - u_h)_{|K} \approx (I_{2h}^{(2)} u_h - u_h)_{|K}.$$

The mesh adaptation is then organized according to the *fixed-fraction* strategy described in Section 4.2. The performance of mesh adaptation on the basis of the weighted error estimator $\eta_\omega(u_h, z_h)$ will be compared to that by the 'energy-norm' error estimator

$$\eta_E := \Big( \sum_{K \in \mathbb{T}_h} \big\{ \|R_h\|_K^2 + h_K \|r_h\|_{\partial K}^2 + \|\nabla \cdot v_h\|_K^2 \big\} \Big)^{1/2},$$

which has already been considered in the Introduction. This estimator can be shown to measure the error in the energy-norm,

$$|||u|||_E := \big( \nu \|\nabla(v - v_h)\|^2 + \|p - p_h\|^2 \big)^{1/2} \approx c_S \eta_E(u_h).$$

Here, the stability constant $c_S$ is related to the constant in the coercivity estimate of the tangent form $a'(u)(\cdot, \cdot)$ taken at the solution $u$,

$$|||z|||_E \leq c_S \sup_{\varphi \in V} \left\{ \frac{A'(u)(z, \varphi)}{|||\varphi|||_E} \right\}, \quad z \in V.$$

This constant is unknown and may be large, so that the estimate cannot be used as a stopping criterion in the mesh adaptation process.

## Numerical results

For the benchmark configuration of Figure 11.1, the drag and lift coefficients $c_{\text{drag}}$ and $c_{\text{lift}}$ are to be computed with an accuracy of, say, 1%. The a priori design of such a mesh is a difficult task. Figure 11.3 shows a collection of meshes where similar ones have actually been used in numerical computations for this benchmark.

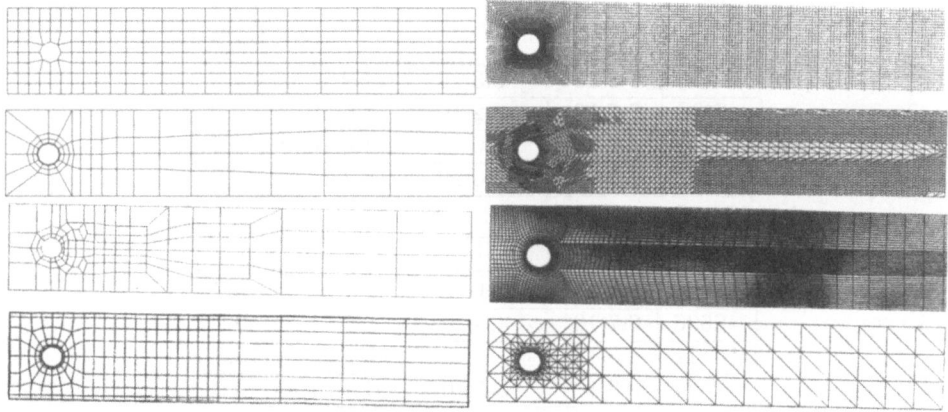

Figure 11.3: *Examples of coarse and fine meshes designed for the benchmark problem 'flow around a cylinder'; see Schäfer and Turek [124]*

Figure 11.4 shows the meshes generated by the weighted error estimator $\eta_\omega$ and the energy-norm error estimator $\eta_E$. The latter induces refinement also down-wind of the cylinder which is not efficient for computing the drag coefficient. The values in Tables 11.1 show a surprisingly good effectivity index for the weighted error estimator $\eta_\omega$. Here, $I_{\text{eff}}$ is defined as before using a *reference solution* which is extrapolated from the finest meshes. In this example the lift coefficient is by two orders of magnitude smaller than the drag coefficient which makes it difficult to achieve a relative error smaller than 1% for both quantities. We also see that the indicator $\Delta\rho$ for the linearization error stays well below that for the discretization error.

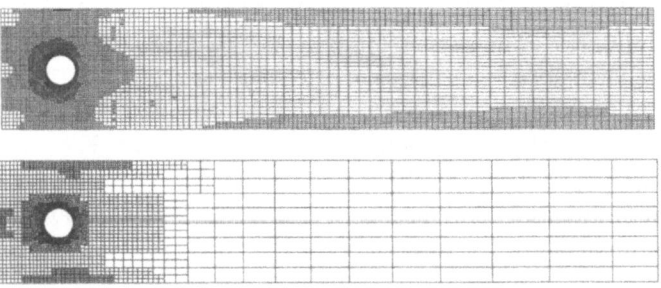

Figure 11.4: *Refined meshes generated by the energy-norm error estimator (top) and by the weighted error estimator (bottom); from Becker [20].*

Table 11.1: *Computation of* $c_{\text{drag}}$ *and* $c_{\text{lift}}$ *on locally refinement meshes; from Becker [20].*

| N | $|J_{\text{drag}}(e)|$ | $|\eta_\omega|$ | $|\Delta\rho|$ | $I_{\text{eff}}$ |
|---|---|---|---|---|
| 458 | $2.2 \cdot 10^{-2}$ | $6.3 \cdot 10^{-2}$ | $1.9 \cdot 10^{-2}$ | 2.8 |
| 1430 | $2.7 \cdot 10^{-3}$ | $3.2 \cdot 10^{-4}$ | $1.6 \cdot 10^{-3}$ | 1.2 |
| 5056 | $4.4 \cdot 10^{-4}$ | $9.0 \cdot 10^{-5}$ | $1.0 \cdot 10^{-4}$ | 0.2 |
| 18644 | $1.1 \cdot 10^{-4}$ | $9.6 \cdot 10^{-5}$ | $5.7 \cdot 10^{-6}$ | 0.9 |

| N | $|J_{\text{lift}}(e)|$ | $|\eta_\omega|$ | $|\Delta\rho|$ | $I_{\text{eff}}$ |
|---|---|---|---|---|
| 708 | $1.6 \cdot 10^{-1}$ | $1.1 \cdot 10^{-1}$ | $4.6 \cdot 10^{-2}$ | 0.7 |
| 2696 | $6.7 \cdot 10^{-2}$ | $3.0 \cdot 10^{-4}$ | $1.2 \cdot 10^{-3}$ | 0.1 |
| 10512 | $1.3 \cdot 10^{-2}$ | $1.1 \cdot 10^{-2}$ | $1.0 \cdot 10^{-4}$ | 0.8 |
| 41504 | $3.3 \cdot 10^{-3}$ | $3.1 \cdot 10^{-3}$ | $6.0 \cdot 10^{-5}$ | 0.9 |

The superiority of the weighted error estimator over the energy-norm error estimator with respect to mesh efficiency is clearly seen in Figure 11.5. The example shown in the Introduction (see Figure 1.4) demonstrates that this effect can be even more pronounced in 3-D configurations.

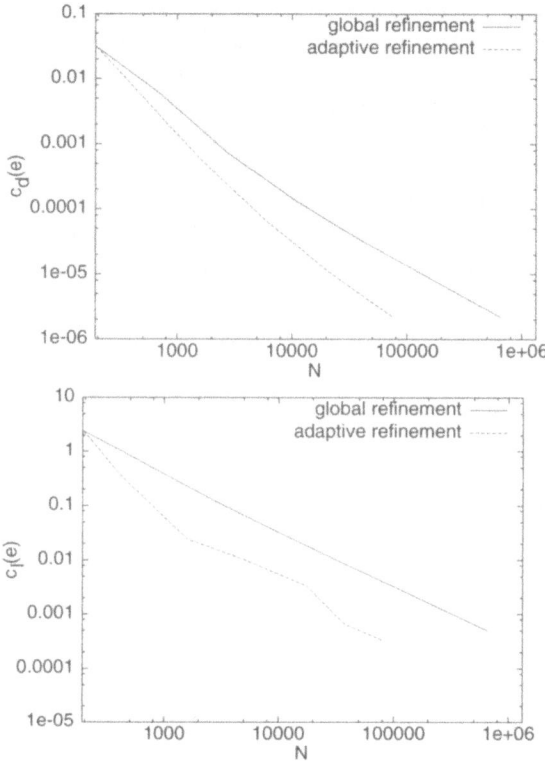

Figure 11.5: *Comparison of mesh efficiencies for the computation of drag (top) and lift (bottom) on globally (solid line) and adaptively (broken line) refined meshes; from Becker [20].*

*Remark* 11.3. Energy-norm-type *a posteriori* error estimates for finite element approximations of the Stokes and Navier-Stokes equations have been derived by Verfürth [130], Bernardi et al. [33], Oden et al. [112], and Ainsworth and Oden [3]. An error analysis based on duality arguments for estimating the $L^2$-norm error was developed by Johnson and Rannacher [90] and Johnson et al. [91]. Another variant of this technique also including functional error estimation based on a posteriori bounds for the error in some energy norm can be found in Machiels et al. [105] and Machiels et al. [106]. In Becker and Rannacher [29] and Becker [18, 19] the application of the DWR approach to drag and lift computation was developed for the cylinder flow benchmark described in Schäfer and Turek [124].

## 11.2   Minimization of drag by boundary control

Next, we consider the optimization of flow, i.e., in the present situation the minimization of the drag by *boundary control* acting as varying negative mean pressures $P_1$ and $P_2$ at the two little openings $\Gamma_1$ and $\Gamma_2$ above and below the cylinder (see Figure 11.6). By these pressure drops the fluid is sucked out of the channel which reduces the drag force on the cylinder until a (local) minimum is reached. Increasing the pressure drop further makes the drag grow again up to some local maximum. Beyond this point the flow state becomes nonstationary.

In order to formulate this optimal control problem in a variational form, we use the same function space $V := H \times L$ for the state variable $u = \{v, p\}$ as above, and $Q = \mathbb{R}^2$ for the control variable $q = \{q_1, q_2\}$. The action of the control is described by the bilinear *control form*

$$b(q, \psi) := -(q_1 n, \psi^v)_{\Gamma_1} - (q_2 n, \psi^v)_{\Gamma_2}, \quad \psi = \{\psi^v, \psi^p\} \in V.$$

With this notation the variational formulation of the optimization problem reads as follows: Find a pair $\{u^{\mathrm{opt}}, q^{\mathrm{opt}}\} \in V \times Q$ for which the linear cost functional

$$J(u, q) := c_{\mathrm{drag}}(u)$$

is a (local) minimum with respect to all pairs $\{u, q\} \in V \times Q$ satisfying the state equation

$$a_\delta(u)(\psi) + b(q, \psi) = 0 \quad \forall \psi \in V. \tag{11.12}$$

Figure 11.6: *Configuration of the drag minimization problem.*

The Galerkin finite element approximation of the optimization problem uses the Lagrangian formalism as described in Chapter 6. For triples $\{u^{\mathrm{opt}}, q^{\mathrm{opt}}, \lambda^{\mathrm{opt}}\} \in \{u^{\mathrm{in}} + V\} \times Q \times V$, we define the Lagrangian functional

$$\mathcal{L}(u, q, \lambda) := J(u, q) - a_\delta(u)(\lambda) - b(q, \lambda),$$

where the adjoint variable $\lambda \in V$ plays the role of a Lagrangian multiplier. We seek stationary points of the Lagrangian functional which are possible (local) solutions of the optimal control problem. These stationary points are determined by the Euler-Lagrange system (*optimality system*):

$$a'(u)(\varphi, \lambda) = J(\varphi) \quad \forall \varphi \in V \tag{11.13}$$

$$b(\chi, \lambda) = 0 \quad \forall \chi \in Q \tag{11.14}$$

$$a(u)(\psi) + b(q, \psi) = 0 \quad \forall \psi \in V \tag{11.15}$$

We note that (11.15) is just the state equation for admissible pairs $\{u, q\}$, (11.13) is the *adjoint problem* associated with the functional $J(\cdot)$, and (11.14) implies an additional boundary condition for the adjoint variable:

$$\int_{\Gamma_i} \lambda^v \cdot n \, ds = 0 \quad (i = 1, 2).$$

The existence of solutions of this system, i.e. of stationary points of the Lagrangian, is made an assumption here. In fact, in the case of 'open' outflow boundary conditions as used here, the question of existence of solutions for realistic data is still an unsettled problem even for the *stationary* Navier-Stokes problem (see Heywood et al. [75]).

For discretizing the Euler-Lagrange system (11.13)–(11.15), we use the same finite element method as above, where now the stabilization is applied independently for the form $a(\cdot)(\cdot)$ and the linearized form $a'(u)(\cdot, \cdot)$ yielding stabilized forms $a_\delta(\cdot)(\cdot)$ and $a'(u)_\delta(\cdot, \cdot)$, respectively. In general, the latter is not identical with the derivative of $a_\delta(\cdot)(\cdot)$,

$$a'(u)_\delta(\cdot, \cdot) \neq a'_\delta(u)(\cdot, \cdot).$$

Therefore, we are in a similar situation as in the preceding section that the general theory of Chapter 8 does not directly apply to the present case. However, the deviation is assumed to be small and we again use the abstract results as guidelines for developing a practically useful adaptation strategy. This will be demonstrated by an example below.

The discrete Euler-Lagrange system determines triples $\{u_h, q_h, \lambda_h\} \in \{u_h^{\mathrm{in}} + V\} \times Q \times V_h$ by

$$a'(u_h)_\delta(\varphi_h, \lambda_h) = J(\varphi_h) \quad \forall \varphi_h \in V_h \tag{11.16}$$

$$b(\chi_h, \lambda_h) = 0 \quad \forall \chi_h \in Q_h \tag{11.17}$$

$$a_\delta(u_h)(\psi_h) + b(q_h, \psi_h) = 0 \quad \forall \psi_h \in V_h. \tag{11.18}$$

Again, the existence of solutions to this nonlinear system is made an assumption. The associated *dual, control*, and *primal residuals* are

$$\rho^*(u_h, \lambda_h)(\cdot) := J(\cdot) - a'_\delta(u_h)(\cdot, \lambda_h),$$
$$\rho^q(\lambda_h)(\cdot) := -b(\cdot, \lambda_h),$$
$$\rho(u_h, q_h)(\cdot) := -a_\delta(u_h)(\cdot) - b(q_h, \cdot).$$

From the general result of Proposition 8.1, we obtain the following a posteriori error representation for the present situation:

$$J(u) - J(u_h) \approx \eta_\omega := \tfrac{1}{2}\rho^*(\lambda_h)(u - \varphi_h) + \tfrac{1}{2}\rho^q(q_h)(q - \chi_h)$$
$$+ \tfrac{1}{2}\rho(u_h)(\lambda - \psi_h), \tag{11.19}$$

for arbitrary $\{\varphi_h, \chi_h, \psi_h\} \in V_h \times Q \times V_h$, where the cubic remainder term is neglected. Evaluating the residuals for the concrete situation as above, we obtain an a posteriori error estimator in terms of the usual cell- and edge-residuals associated to the equations (11.13)–(11.15). The details of this estimator are not stated here. On this basis the mesh adaptation for the discrete optimality system (11.16)–(11.18) uses again the fixed-fraction strategy.

### Numerical results

We present some results obtained for the minimization of the drag coefficients by boundary control as shown in Figure 11.6. The data is chosen such that $Re = \bar{U}^2 D / \nu = 40$ for the uncontroled flow. In Table 11.2, we compare the values of the drag coefficient on optimized meshes as shown in Figure 11.7 with results obtained on uniformly refined meshes. We see that a significant reduction in the dimension of the discrete model is possible by using appropriately adapted meshes.

Table 11.2: *Uniform refinement (left) versus adaptive refinement (right); from Becker [20].*

| Uniform refinement | | Adaptive refinement | |
|---|---|---|---|
| $N$ | $J_{\mathrm{drag}}$ | $N$ | $J_{\mathrm{drag}}$ |
| 10, 512 | 3.31 | 1, 572 | 3.29 |
| 41, 504 | 3.21 | 4, 264 | 3.17 |
| 164, 928 | 3.12 | 11, 146 | 3.12 |

Figure 11.7 shows streamline plots of the uncontroled $(q = 0)$ and the controlled $(q = q^{\mathrm{opt}})$ solution and a corresponding 'optimal' mesh. The locally refined mesh produced by the adaptive algorithm seems to contradict intuition since the recirculation behind the cylinder is not resolved. However, due to the particular structure of the optimal velocity field (most of the flow leaves the domain at the control boundary), it might be clear that this recirculation does not significantly influence the cost functional. Instead, a strong local refinement near the cylinder, where the cost functional is evaluated, as well as near the control boundary is produced. However, we have to worry about the dynamic stability of this solution, in view of its rich vortex structure. We will investigate this aspect in the next section.

The flow field $v_h^{\mathrm{opt}}$ of the drag-minimal solution computed on an adapted mesh does not represent the 'true' flow in the downwind region of the channel, i.e., it is not admissible in the strict sense for the optimization problem. This is seen by comparing it with the flow field $\tilde{v}_h^{\mathrm{opt}}$ computed by post-processing on a globally refined mesh $\mathbb{T}_h^{\mathrm{glob}}$,

$$a_\delta(\tilde{u}_h^{\mathrm{opt}})(\psi_h) = -b(q_h^{\mathrm{opt}}, \psi_h) \quad \forall \psi_h \in V_h^{\mathrm{glob}}; \tag{11.20}$$

see Figure 11.8. As predicted by the theory, the drag coefficient computed from this post-processed solution is (almost) the same as the 'optimal' one obtained on the adapted mesh.

Figure 11.7: *Adapted mesh for the drag minimization (top) and streamline plots of the uncontroled flow (middle) and controled flow (bottom) computed on these meshes; from Becker [20].*

Figure 11.8: *Streamline plots of the velocity field of the 'minimal drag' solution (base solution) computed by post-processing on a globally refined mesh; from Heuveline and Rannacher [78].*

## 11.3  Stability analysis for stationary flow

The drag minimization described above has been carried out by a 'stationary' flow solver, i.e., the nonlinear optimality system (11.13)–(11.15) is solved by a quasi-Newton iteration. It is well known that by such an approach solutions $\hat{u}$ can be obtained which are physically unstable. Here, 'unstable' means that there are small perturbations $\delta\hat{u}$ such that the nonstationary solution $\{u(t) \in V, t \geq 0\}$, starting from $u(0) = \hat{u} + \delta\hat{u}$, and satisfying the nonstationary Navier-Stokes equations

$$\partial_t v - \nu \Delta v + v \cdot \nabla v + \nabla p = 0, \quad \nabla \cdot v = 0, \tag{11.21}$$

does not tend back to $\hat{u}$ as $t \to \infty$. For example, for the benchmark configuration of Figure 11.1, stationary solutions can be obtained by a Newton method on relevant meshes for Reynolds numbers up to Re $\approx$ 100. However, this flow is known to turn nonstationary for Reynolds numbers around Re $\approx$ 75. Figure 11.9 shows a sequence of streamline plots of computed stationary flows around the cylinder for Re = 40 (determined by the inflow velocity) but with increasing equal pressure drop imposed at the two outlets above and below the cylinder. It is hardly possible to decide just from the pattern of these flows whether they are stable or unstable.

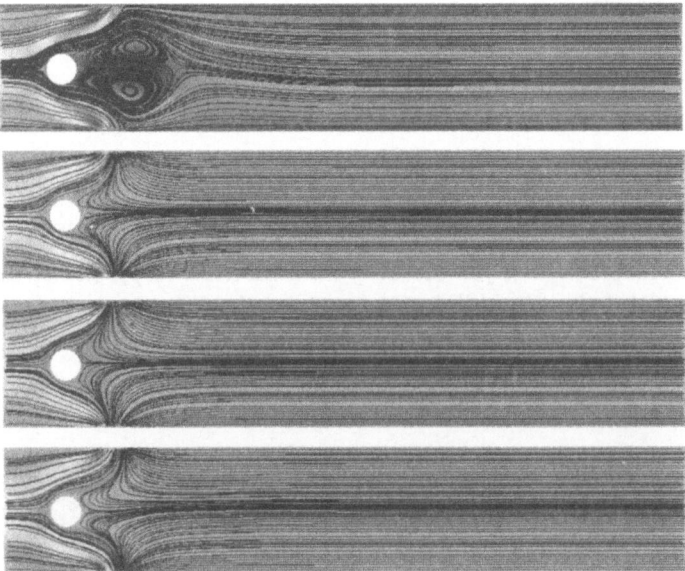

Figure 11.9: *Streamline plots of stationary flows around a cylinder with imposed equal pressure drop of $\Delta P = 0.10,\ 0.34,\ 0.50,\ 1.00$ at the outlets; from Heuveline & Rannacher [78].*

Let the base solution $\hat{u} = \{\hat{v}, \hat{p}\}$ be determined by the stationary Navier-Stokes system in weak form

$$a(\hat{u})(\psi) = -b(q, \psi) \quad \forall \psi \in V, \tag{11.22}$$

with fixed 'force term' $b(q, \varphi)$ such as considered in the drag minimization problem. In Section 7.3, we have discussed that the instability of this base solution can be detected by investigating the nonsymmetric *stability eigenvalue problem* (*linearized stability analysis*)

$$\mathcal{A}'(\hat{u})u := \begin{bmatrix} -\nu \Delta v + \hat{v}\cdot\nabla v + v\cdot\nabla\hat{v} + \nabla p \\ \nabla\cdot v \end{bmatrix} = \lambda \begin{bmatrix} v \\ 0 \end{bmatrix} =: \lambda \mathcal{M} u. \tag{11.23}$$

If there is a *critical* eigenvalue $\lambda^{\mathrm{crit}}$ with negative real part, then $\hat{u}$ is unstable.

*Remark 11.4.* The *nonlinear stability analysis* considers the symmetric eigenvalue problem

$$\mathcal{A}'(\hat{u})u := \begin{bmatrix} -\nu \Delta v + \frac{1}{2}(\nabla\hat{v}+\nabla\hat{v}^T)v + \nabla p \\ \nabla\cdot v \end{bmatrix} = \lambda \begin{bmatrix} v \\ 0 \end{bmatrix} =: \lambda \mathcal{M} u.$$

If all its eigenvalues have non-negative real part, then the reference flow $\hat{u}$ is stable. However, this strong criterion is not practical for complex flows such as the benchmark considered here, since it guarantees *unconditional* stability, i.e. stability with respect to arbitrarily large perturbations.

Using the sesquilinear forms

$$a'(\hat{u})(\varphi, \psi) := \nu(\nabla\varphi^v, \nabla\psi^v) + (\hat{v}\cdot\nabla\varphi^v, \psi^v) + (\varphi^v\cdot\nabla\hat{v}, \psi^v) - (\varphi^p, \nabla\cdot\psi^v) \\ - (\psi^p, \nabla\cdot\varphi^v),$$

and $m(\varphi, \psi) := (\varphi^v, \psi^v)$, the weak forms of the primal and dual eigenvalue problem associated with (11.23) read as follows:

$$a'(\hat{u})(u, \psi) = \lambda\, m(u, \psi) \qquad \forall \psi \in V, \tag{11.24}$$
$$a'(\hat{u})(\varphi, u^*) = \lambda\, m(\varphi, u^*) \qquad \forall \varphi \in V, \tag{11.25}$$

with the normalization $m(u, u) = m(u, u^*) = 1$. In the following the symbol $\lambda$ will be used to denote eigenvalues, which is not to be confused with its use for the *adjoint* variable in the last section. Note that in setting up the stability eigenvalue problem, all linear nonhomogeneous terms such as volume force terms and Dirichlet or Neumann boundary data vanish upon differentiation of the form $a(\cdot)(\cdot)$. That is why the boundary term $b(q, \cdot)$ does not occur in the eigenvalue problem.

*Remark 11.5.* In the case that $u$ and $u^*$ are $m$-orthogonal, i.e. $m(u, u^*) = 0$, the boundary value problem, for fixed $\{u, \lambda\}$,

$$a'(\hat{u})(\tilde{u}, \psi) - \lambda\, m(\tilde{u}, \psi) = m(u, \psi) \qquad \forall \psi \in V,$$

has a solution $\tilde{u} \in V$ which is a so-called *generalized eigenfunction*. Hence, the eigenvalue $\lambda^{\mathrm{crit}}$ is deficient. i.e. its algebraic multiplicity is larger than the geometric one.

For the discretization of the eigenvalue problem (11.23), we employ an analogous finite element method as used for computing the approximate base solution $\hat{u}_h = \{\hat{v}_h, \hat{p}_h\}$,

$$a(\hat{u}_h)(\psi_h) = -b(q, \psi_h) = 0 \quad \forall \psi_h \in V_h. \tag{11.26}$$

Introducing the stabilized sesquilinear form

$$a'(\hat{v}_h)_\delta(u_h, \psi_h) := a'(\hat{u})(u_h, \psi) + (\mathcal{A}'(\hat{u})u - \lambda_h v, \mathcal{S}(\hat{u})\psi)_\delta$$

the discrete primal and dual eigenvalue problems for $u_h, u_h^* \in V_h$, $\lambda_h \in \mathbb{C}$ read as follows:

$$a'(\hat{u}_h)_\delta(u_h, \psi_h) = \lambda_h \, m(u_h, \psi_h) \qquad \forall \psi_h \in V_h \tag{11.27}$$

$$a'(\hat{u}_h)_\delta(\varphi_h, u_h^*) = \lambda_h \, m(\varphi_h, u_h^*) \qquad \forall \varphi_h \in V_h, \tag{11.28}$$

with the normalization condition $m(u_h, u_h) = m(u_h, u_h^*) = 1$. The singular case $m(u_h, u_h^*) = 0$ will rarely occur on the discrete level. It would indicate that the limit eigenvalue is close to being deficient.

Then, from the general result of Proposition 7.9, neglecting higher-order remainder terms, we have the following a posteriori error representation:

$$\begin{aligned} \lambda - \lambda_h \approx \eta_\omega := {} & \tfrac{1}{2}\hat{\rho}(\hat{u}_h)(\hat{u}^* - \hat{\psi}_h) + \tfrac{1}{2}\hat{\rho}^*(\hat{u}_h^*)(\hat{u} - \hat{\varphi}_h) \\ & + \tfrac{1}{2}\rho(u_h, \lambda_h)(u^* - \psi_h) + \tfrac{1}{2}\rho^*(u_h^*, \lambda_h)(u - \varphi_h), \end{aligned} \tag{11.29}$$

for arbitrary $\hat{\varphi}_h \in u_h^{\mathrm{in}} + V_h$ and $\hat{\psi}_h, \varphi_h, \psi_h \in V_h$. The primal and dual residuals for the approximation of the base solution are

$$\hat{\rho}(\hat{u}_h)(\cdot) := -a_\delta(\hat{u}_h)(\cdot),$$

$$\hat{\rho}^*(\hat{u}_h^*)(\cdot) := -a_\delta''(\hat{u})(\cdot, u_h, u_h^*) - \tilde{a}_\delta'(\hat{u}_h)(\cdot, \hat{u}_h^*),$$

and those for the eigenvalue approximation

$$\rho(u_h, \lambda_h)(\cdot) := \lambda_h \, m(u_h, \cdot) - \tilde{a}'(\hat{u}_h)_\delta(u_h, \cdot),$$

$$\rho^*(u_h^*, \lambda_h)(\cdot) := \lambda_h \, m(\cdot, u_h^*) - \tilde{a}'(\hat{u}_h)_\delta(\cdot, u_h^*).$$

The underlying meshes in computing $\hat{u}_h$ may have been adapted towards the economical computation of the drag coefficient and do not provide a uniformly good approximation of $\hat{u}$. Hence the question arises whether the accuracy of $\hat{u}_h$ in approximating $\hat{u}$ is sufficient to obtain reliable results for the stability of the latter one, and whether we are told by our a posteriori error estimator if this is the case or not. This will be investigated by numerical tests below.

## Numerical results

The stability eigenvalue problem (11.23) has been solved for several values of the control $q$ with mesh adaptation on the basis of the weighted error estimator $\eta_\omega^\lambda$. Figure 11.10 displays the real and imaginary parts of the critical eigenvalue $\lambda^{\text{crit}}$ as function of the control pressure drop. The 'optimal' pressure drop for the stationary model is around $\Delta P = 0.5$ and the corresponding flow state turns out to be unstable. Figure 11.11 shows two meshes which are adapted for computing the drag-minimal base solution and for solving the stability eigenvalue problem.

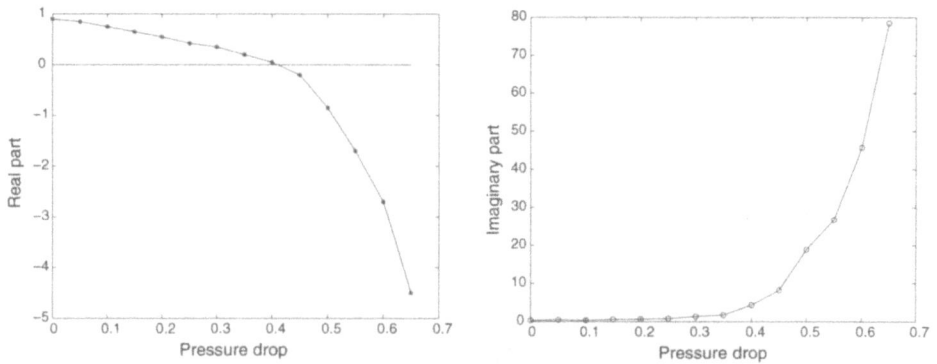

Figure 11.10: *Real and imaginary parts of the critical eigenvalue as function of the pressure drop; from Heuveline and Rannacher [78].*

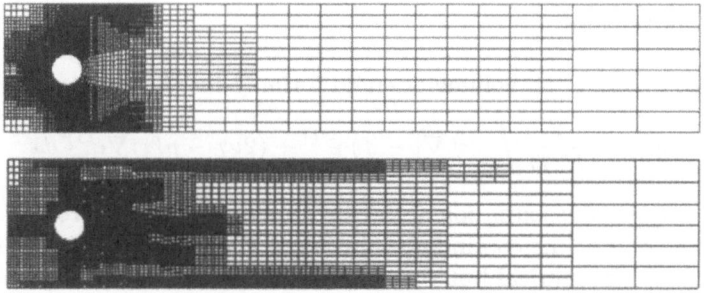

Figure 11.11: *Meshes obtained by the error estimators for the drag (top) and the eigenvalue computation (bottom); from Heuveline and Rannacher [78].*

In Figure 11.12 we show the size of the two components in the error estimator (11.29), measuring the error in the approximation of the base solution and the error in the eigenvalue approximation. On coarse meshes the error in approximating the base solution, i.e. the coefficient in the stability eigenvalue problem, dominates, while under further mesh refinement the error becomes smaller than that of the eigenvalue approximation. In particular, we see that on the finest mesh which is adapted in accordance to the drag minimization process, we can reliably predict the stability, in this case actually the instability, of the corresponding flow state.

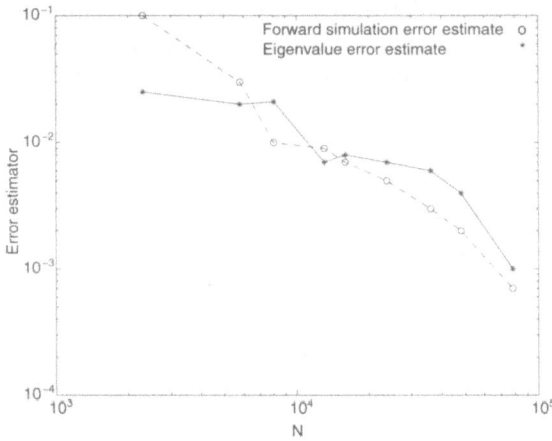

Figure 11.12: *The size of the two components of the error estimator $\eta_\omega^\omega$, i.e. the errors in the base solution and the eigenvalue approximation; from Heuveline and Rannacher [78].*

## 11.4 Exercises

*Exercise* 11.1. Derive the volume-oriented formula for the drag coefficient,

$$c_{\mathrm{drag}} = \frac{2}{\bar{U}^2 D} \int_{\partial\Omega} (v\cdot\nabla v - f)\cdot\bar{e}^{(1)} + (2\nu\tau - pI) : \nabla\bar{e}^{(1)} \, dx,$$

where $\bar{e}^{(1)}$ is a suitable extension of the directional vector $e^{(1)}$ from $S$ to $\Omega$ with support along $S$. For the continuous solution $u$ the surface- and volume-oriented representations are identical but they differ for the discrete solution $u_h$.

*Exercise* 11.2. Verify the following explicit form of the dual cell and edge residuals:

$$R_{h|K}^* = j + \nu\Delta z_h^v + (\nabla\cdot v_h)z_h^v + v_h\cdot\nabla z_h^v - (\nabla v_h)z_h^v - \nabla z_h^p$$

$$r_{h|\Gamma}^* = \begin{cases} \frac{1}{2}[\nu n\partial_n z_h^v + (v_h\cdot n)z_h^v - nz_h^p], & \text{if } \Gamma \not\subset \partial\Omega, \\ -\nu n\partial_n z_h^v - (v_h\cdot n)z_h^v + nz_h^p, & \text{if } \Gamma \subset \partial\Omega_{out}, \\ 0, & \text{otherwise.} \end{cases}$$

occurring in the a posteriori error estimator $\eta_\omega$ for the drag coefficient.

# Chapter 12

# Miscellaneous and Open Problems

The material presented in these Lecture Notes demonstrates that duality-based error estimation as realized in the DWR method can be applied, in principle, to all problems posed in variational form, even if their regularity properties do not match all assumptions. The additional work for the adaptive component of the solution process is relatively small. In fact, for nonlinear problems the evaluation of the a posteriori error estimates amounts to about the equivalent of *one* extra step within the outer Newton iteration on the current mesh. In this case, the extra work for mesh adaptation usually makes up 5−25% of the total work on the *optimized* mesh. However, the implementation may appear difficult when existing software components like mesh generators, multigrid solvers, etc., cannot be used directly.

There are still many open problems with respect to the practical realization of the DWR method. However, the effort to be invested is worthwhile even for very complex problems, because once the method works, the possible gain in accuracy and solution efficiency can be significant.

## 12.1   Some historical remarks

Though the roots of the variational finite element method date back to the beginning of the last century, the first real 'finite element paper' is that of Courant [47] where the 'Courant-triangle' is described. The rigorous mathematical *a priori* error analysis of the method began only in the sixties independently with papers by several authors from China, Russia, the USA, and Europe; the most influential one was probably that of Zlámal [138]. For more detailed historical surveys of the finite element method see Oden [109] and Babuška and Strouboulis [12].

The *a posteriori* error analysis for finite element methods was initiated by the pioneering work of Babuška and Rheinboldt [9, 10] and has then been further developed by Ladeveze and Leguillon [100], Bank and Weiser [17], Babuška and Miller [8], and Ainsworth and Oden [1], to mention only a few of the most influential papers. For discussions and further references, we refer to the surveys by Verfürth [132], and Ainsworth and Oden [2], and to the recent book of Babuška and Strouboulis [12]. Most of this work is concerned with elliptic problems and error estimation with respect to natural *energy norms* or other related global norms. This seems rather generic as it is directly based on the variational formulation of the underlying problem and allows to use its natural coercivity properties. However, as we have seen, in most applications the error in the energy norm does not provide useful bounds on the errors in the quantities of real physical interest. This goal is accomplished by duality-based approaches such as the DWR method described in these Lecture Notes.

The use of 'duality' in *a posteriori* error estimation goes back to ideas of Babuška and Miller [5, 6, 7] in the context of post-processing of 'quantities of physical interest' in elliptic model problems. Duality arguments as used in the DWR method have first played a central role in the *a priori* error analysis of the finite element method in order to derive optimal-order error estimates in norms other than the natural energy norm. In that context, the approach is known as the 'Aubin-Nitsche trick', see Aubin [4] and Nitsche [107]. Since then, this idea has been used successfully for all kinds of problems of elliptic, parabolic as well as hyperbolic type. However, for a long time the *a posteriori* error analysis was limited to linear elliptic problems and concentrated on error estimation in the energy norm. The use of duality arguments in this context was first systematically developed by Johnson [84], Eriksson and Johnson [51, 52], and their collaborators extending also to time-dependent problems (see also Johnson [87] and the survey paper by Eriksson et al. [50]). Here, the main focus still was on estimating the error in global norms and stability constants for the dual problems were mostly derived by analytical arguments. In Becker and Rannacher [29, 30] this approach was developed into a computation-based feedback method, the *DWR method*, for computing local quantities of interest (see the survey papers mentioned in the Preface.) About the same time, following ideas by Babuška and Miller [5, 6, 7], related techniques based on energy-norm error estimation have been proposed by Paraschivoiu and Patera [113], Machiels et al. [105], and Prudhomme and Oden [110].

## 12.2   Current developments

The DWR method and related approaches are currently being developed further for various classes of problems and methods. We list some of the most important areas of research.

- *3-D applications:* The full power of the DWR method is expected to be seen in applications to 3-D problems. Those are currently developed in the context of Navier-Stokes equations for chemically reactive flows such as laminar methane combustion (see Braack et al. [40]), and astrophysical models involving radiative transfer (see Kanschat [93, 94], Führer and Kanschat [62], and Richling et al. [122]).

- *Multi-physics and multi-scale problems:* The DWR method allows the systematic treatment of rather complex systems involving various physical mechanisms and spatial or temporal scales. A typical example is seen in chemically reacting flow where the chemical reaction usually induces much faster time scales and smaller spatial scales compared to those of the flow. The application of the DWR method to problems of this type is described in Braack [39], Becker et al. [26, 25], Braack and Rannacher [42], Hebeker and Rannacher [74], and Waguet [135].

- *Adaptivity in optimal control:* Most numerical simulation eventually is optimization. Complex multidimensional optimal control and parameter identification problems constitute highly demanding computational tasks. Goal-oriented model reduction by adaptive discretization has the potential of facilitating large-scale optimization problems in structural and fluid mechanics, such as for example, minimization of drag or control of flow-induced structural vibrations, as well as in the estimation of distributed parameters in PDE. The use of mesh adaptation techniques such as the DWR method for this purpose has just begun (see Becker [21], Bangerth [15], and Becker and Vexler [32]).

- *Model adaptivity:* The concept of *a posteriori* error control for single quantities of interest via duality may also be applicable to other situations when a full model, such as a differential equation, is reduced by projection to a subproblem, such as a finite element model. Model reduction within scales of hierarchical sub-models is a recent development in structural as well as fluid mechanics (see Babuška and Schwab [11], Stein and Ohnimus [127], and Hughes et al. [81]). Quantitative error control within these techniques by computational means like the DWR method seems to be a promising idea (see Oden and Prudhomme [111] and Braack and Ern [41]).

- *Open areas for applications:* There are several very promising and almost untouched areas for the application of the DWR method such as, e.g., electro-magnetics, semi-conductor theory, porous media flow, and fluid-structure interaction.

- *'Non-standard' finite element methods:* The DWR method can also be used in the context of 'non-standard' finite element methods such as *non-conforming* and *mixed* methods (see Dunne [49]) as well as the 'least-squares' stabilized cG-FEM for transport-dominated problems (see Führer [61], Führer and Rannacher [63], Houston et al. [79]). A complete analysis is available for the dG-FEM for elliptic problems (see Kanschat and Rannacher [95]), and for hyperbolic problems such as the Euler equations in fluid mechanics (see Hartmann [70, 71] and Hartmann and Houston [73, 72]). For completeness, we also mention the earlier works of Eriksson and Johnson [53], Hansbo and Johnson [68], Johnson and Szepessy [92] in which duality-based techniques are used for deriving energy-norm-type error estimates involving global stability constants.

## 12.3   Open problems

The performance of the DWR method has been demonstrated for several linear and nonlinear model problems, mainly from solid and fluid mechanics. In this context there are several theoretical and practical problems which have so far not been satisfactorily treated. Below, we list some of them together with relevant literature:

- *How to use the DWR method for multidimensional time-dependent problems?* Rigorous error control in the space-time frame requires us to solve a space-time dual problem. Especially for nonlinear problems, this may be prohibitive with respect to storage space and computing time. The question is how to exploit the option of solving on coarser meshes only and that of data compression. The realization of these concepts for practical problems beyond simple model situations is still in an immature state (see Becker [21]).

- *How to apply the DWR method in the context of the hp-method?* Here, the goal is to design a strategy for simultaneous adaptation of mesh size $h$ and polynomial degree $p$ which is rigorous also with respect to the $p$-adaptation.

- *How to organize anisotropic mesh refinement?* The rigorous extension of the DWR concept for generating solution-adapted *anisotropic* meshes, either by simple cell stretching or by more sophisticated mesh reorientation, is still to be developed; see the discussion in Section 4.4. In the context of 'global' error estimation with respect to energy norm and $L^2$ norm a posteriori error estimates on anisotropic meshes have been derived, e.g., in Siebert [126] and Kunert [97, 98].

- *How to effectively control the error caused by 'variational crimes'?* These are deviations from the pure Galerkin method such as numerical integration, boundary approximation, cutting off unbounded domains, transport stabilization, etc. (see Rannacher [115], Dörfler and Rumpf [48], Zamni [136]).

- *How to control the 'algebraic' solution errors?* These are unavoidable using iterative solvers such as Newton's method, Krylov-space methods, multigrid methods, etc. (see Becker et al. [27], Becker [19], Larson and Niklasson [102], and Becker and Braack [23]).

- *How to apply the DWR method to non-variational problems?* Residual-based methods for *a posteriori* error estimation like the DWR method rely on the variational formulation and the Galerkin orthogonality property of the finite element scheme. This allows us to locally extract additional powers of the mesh size, leading to sensitivity factors of the form $h_K^2 \|\nabla^2 z\|_K$ . Other 'non-variational' discretizations such as the finite volume method usually have a different error behavior, governed by sensitivity factors like $h_K \|\nabla z\|_K$ . This may be seen by reinterpreting these discretizations as perturbed finite element schemes obtained by evaluating local integrals by special low-order quadrature rules. An *a posteriori* error analysis for finite volume schemes exploiting superapproximation properties has been developed by Giles [64] and Giles and Pierce [65].

- *How to solve the technical problems discussed in Chapter 5 for providing theoretical support for the DWR method?*

# Acknowledgments

We wish to thank several of the present and past members of the Numerical Analysis Group at Heidelberg for important contributions to these Lecture Notes, particularly R. Becker, M. Braack, R. Hartmann, G. Kanschat, H. Kapp, T. Richter, F.-T. Suttmeier, C. Waguet, and B. Vexler. They have provided various results for particular cases and much of the computational material.

Finally, we acknowledge the financial support by the Deutsche Forschungsgemeinschaft (DFG) via the SFB 359 'Reactive Flow, Diffusion and Transport' and the Graduiertenkolleg 'Modelling and Scientific Computing in Mathematics and the Natural Sciences' at the University of Heidelberg, and the support and hospitality of the Department Mathematik at the ETH Zürich during the summer term 2002.

# Appendix A

# Solutions of exercises

In the following, we present sample solutions together with some further discussions for the theoretical and practical exercises posed in the preceding chapters.

## Chapter 2

**Exercise 2.1** For the present situation, we determine

$$K(T) \approx (1-T)^{-2}, \qquad |\tau_n^0(U)| \approx \sup_{I_n} |u''| \approx (1-t_n)^{-3},$$

and therefore $h_n \approx (1-T)(1-t_n)^{3/2} N^{-1/2} \, \mathrm{TOL}^{1/2}$. This yields

$$N = \sum_{n=1}^{N} h_n h_n^{-1} \approx N^{1/2}(1-T)^{-1}\mathrm{TOL}^{-1/2} \sum_{n=1}^{N} h_n (1-t_n)^{-3/2}$$
$$\approx N^{1/2}(1-T)^{-3/2}\mathrm{TOL}^{-1/2},$$

and, finally, $N \approx (1-T)^{-3}\mathrm{TOL}^{-1}$.

**Exercise 2.2** This time, we have $\rho_n \approx \sup_{I_n} |u'| \approx (1-t_n)^{-2}$,

$$\omega_n = \int_{I_n} |z'| \, dt = 2(1-T)^{-2} \int_{I_n} (1-t) \, dt \approx h_n (1-t_n)(1-T)^{-2},$$

and, analogously to Exercise 2.1, $h_n \approx (1-t_n)^{1/2}(1-T)N^{-1/2}\mathrm{TOL}^{1/2}$. Thus,

$$N = \sum_{n=1}^{N} h_n h_n^{-1} \approx N^{1/2}(1-T)^{-1}\mathrm{TOL}^{-1/2} \sum_{n=1}^{N} h_n (1-t_n)^{-1/2}$$
$$\approx N^{1/2}(1-T)^{-1}\mathrm{TOL}^{-1/2},$$

and, finally, $N \approx (1-T)^{-2}\mathrm{TOL}^{-1}$.

**Exercise 2.3** For the cG(1) method the dual problem in variational form looks similar to that for the dG(0) method, but without the jump terms:

$$\int_I (\varphi, -z' - B^* z)\, dt + (\varphi(T), z(T)) = (\varphi(T), e_N \|e_N\|^{-1}).$$

Take $\varphi := e$, use integration by parts and Galerkin orthogonality, to obtain

$$\|e_N\| = \int_I (e, -z' - B^* z)\, dt + (e(T), z(T)) = \int_I (e' - Be, z)\, dt + (e(0), z(0))$$

$$= \int_I (e' - f(u) + f(U), z)\, dt = \int_I (f(U) - \overline{f(U)}, z - \bar{z})\, dt,$$

where $\overline{f(U)}$ denotes the interval-wise mean value of $f(U)$. From this, we obtain

$$\|e_N\| \leq c_I \sum_{n=1}^{N} h_n \left( \sup_{I_n} \|f(U) - \overline{f(U)}\| \right) \left( \int_{I_n} \|z'\|\, ds \right) =: c_I \sum_{n=1}^{N} h_n \rho_n \omega_n.$$

**Exercise 2.4 (practical exercise)** The actually computed relative error $|u(T) - U(T)|/|u(T)|$ is compared in Figure A.1 for two different end times $T = 0.95$ and $T = 0.999$, and global refinement as well as the mesh-width formulas used in Exercises 2.1 and 2.2. All three choices lead to an error that is proportional to $N^{-1}$, for all values of the end time $T$.

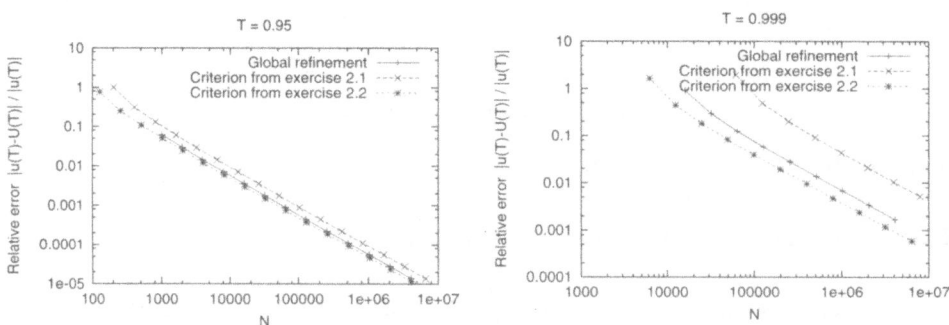

Figure A.1: *Error for various mesh width choices, for end times $T = 0.95$ (left) and $T = 0.999$ (right).*

    Comparing the mesh-width criteria against each other shows that the implicit mesh choice strategy developed from the finite difference point of view is even worse than a globally uniform mesh. On the other hand, the strategy based on the finite element formulation is better than global refinement, with a small but growing margin for end times that are not too close to the critical value $t = 1$. The results do not significantly change if the truncation error and the residuals are approximated numerically, instead of using knowledge of the exact solution as in Exercises 2.1 and 2.2.

If the mesh-width formulas are to be used as stopping criteria, then the actually computed error should not be too far away from the prescribed tolerance for which the respective mesh was made. The left panel of Figure A.2 shows that for the criterion used in Exercise 2.2, these values coincide very well, while the criterion used in Exercise 2.1 produces an error that is actually much better than the prescribed tolerance. This, of course, is due to the fact that in its derivation we have used the worst-case stability constant $K(T)$ rather then the temporally varying dual solution. This overestimation becomes worse as $T \to 1$, and using the tolerance in the mesh choice formula as a stopping criterion will lead to much unnecessary numerical work.

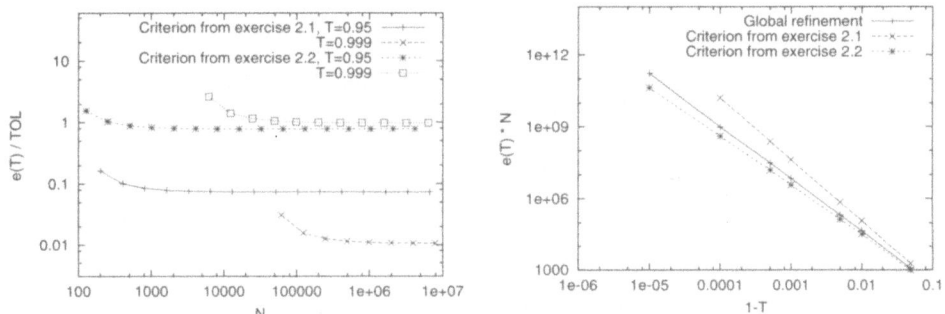

Figure A.2: *Ratio of true (computed) error and prescribed tolerance (left); error $e(T)N$ as a function of $1-T$ (right).*

In order to check the behavior of the errors as $T \to 1$, we show the behavior of $e(T) \cdot N$ in the right panel of Figure A.2 as a function of $1-T$. The theoretical predictions that $e(T) \cdot N$ be proportional to $|\log(1-T)|(1-T)^{-2}$ (uniform mesh), $(1-T)^{-3}$ (mesh choice after Exercise 2.1), or $(1-T)^{-2}$ (mesh choice after Exercise 2.2) are well confirmed. Compared to global refinement, adaptive mesh refinement thus only gains if we move $T$ very close to 1, for this simple model example.

As a final comparison, Figure A.3 shows the errors at the end time when using the second order cG(1) method with global refinement, and the first order dG(0) method with the optimal mesh width strategy as used in Exercise 2.2. The superiority of the former is striking, in particular as the numerical effort is not much higher. Optimal mesh choice would further improve this method.

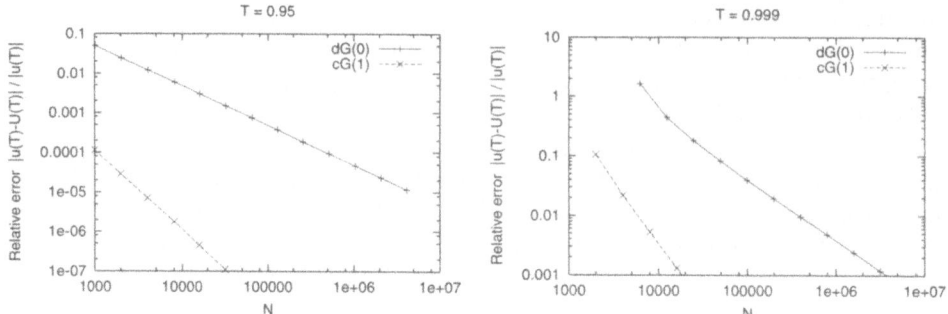

Figure A.3: *Comparison of the cG(1) and the dG(0) method for $T = 0.95$ and $T = 0.999$.*

# Chapter 3

**Exercise 3.1** The approach based on energy error estimates for primal and dual solutions is attractive since such estimates are well known and provide a high degree of accuracy. However, consider the point evaluation of the derivative error

$$J(e) = \partial_1 e(a), \quad a \in \Omega.$$

On uniformly refined meshes, there holds $|J(e)| = \mathcal{O}(h)$, but for $u \in H^2(\Omega)$, and for the singular dual solution $z$, we only have

$$\|\nabla e\| = \mathcal{O}(h), \quad \|\nabla e^*\| = \mathcal{O}(h^{-1}).$$

On the other hand, on 'optimally' refined meshes,

$$|J(e)| = \mathcal{O}(\mathrm{TOL}),$$

and

$$\|\nabla e\| = \mathcal{O}(\mathrm{TOL}^{1/2}), \quad \|\nabla e^*\| = \mathcal{O}(1).$$

In both cases the energy-norm-error-based estimate is of sub-optimal order. Note that we also cannot extract refinement indicators from the estimate, since it cannot be localizes, i.e., the estimate

$$(\nabla e, \nabla e^*) \leq \sum_{K \in \mathbb{T}_h} h_K \|\nabla^2 u\|_K h_K \|\nabla^2 z\|_K$$

does not hold.

**Exercise 3.2** The basic idea is to first derive an estimate for the point error at an arbitrary but fixed point $a$, and later try to use a priori estimates that remove all references to this particular point. For an arbitrary point $a \in \Omega$, use the output functional

$$J_\varepsilon(e) := |B_\varepsilon(a)|^{-1} \int_{B_\varepsilon(a)} e \, dx,$$

with $\varepsilon = \text{TOL}$, which results in

$$|e(a)| \le |J_\varepsilon(e)| + \mathcal{O}(\text{TOL}).$$

For the solution (regularized Green function) of the corresponding dual problem

$$-\Delta z = \delta_\varepsilon^a \quad \text{in } \Omega, \qquad z_{|\partial\Omega} = 0,$$

with $\delta_\varepsilon^a(x) \equiv |B_\varepsilon(a)|^{-1}$ for $x \in B_\varepsilon(a)$, and $\delta_\varepsilon^a \equiv 0$ elsewhere, we have the a priori bound (see, e.g., Ciarlet [46])

$$\|r \nabla^2 z\| \le c\|r^{-1}\| \le c|\log(\text{TOL})|^{1/2},$$

where $r(x) := (|x - a|^2 + \varepsilon^2)^{1/2}$. Note that the last result does not contain references to the point $a$ any more. Now, we estimate as follows:

$$|J_\varepsilon(e)| \le c_I \sum_{K \in \mathbb{T}_h} h_K^2 \rho_K \|\nabla^2 z\|_K$$

$$\le c_I \Big( \sum_{K \in \mathbb{T}_h} h_K^4 \rho_K^2 r_K^{-2} \Big)^{1/2} \Big( \sum_{K \in \mathbb{T}_h} r_K^2 \|\nabla^2 z\|_K^2 \Big)^{1/2}$$

$$\le cc_I \max_{K \in \mathbb{T}_h} \{h_K \rho_K\} \|r^{-1}\| \, \|r \nabla^2 z\| \le cc_I |\log(\text{TOL})| \max_{K \in \mathbb{T}_h} \{h_K \rho_K\}.$$

Hence, the refinement criterion

$$h_K \rho_K \le \frac{\text{TOL}}{cc_I |\log(\text{TOL})|}$$

yields the required accuracy for the $L^\infty$-norm error.

**Exercise 3.3 (practical exercise)** a) Figure A.4 shows two optimal meshes generated by the weighted error estimator for the functionals $J(u) = u(a)$ and $J(u) = \partial_1 u(a)$, respectively, with $a = (0.75, 0.75)$. Figure A.5 shows the quality of error estimation on a sequence of these meshes, where for the evaluation of the error representation the dual solution was approximated by piecewise quadratic finite elements. As can be seen, the error is estimated very well.

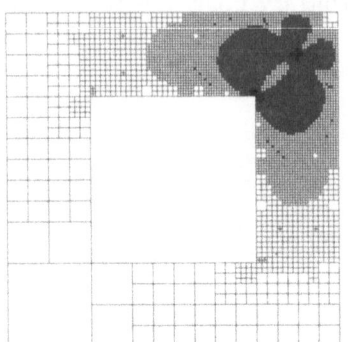 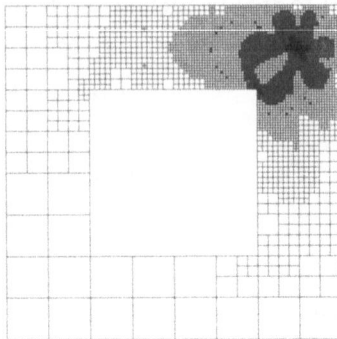

Figure A.4: *Meshes after 7 refinement cycles for the point value $u(a)$ (left) and the point derivative $\partial_1 u(a)$ (right).*

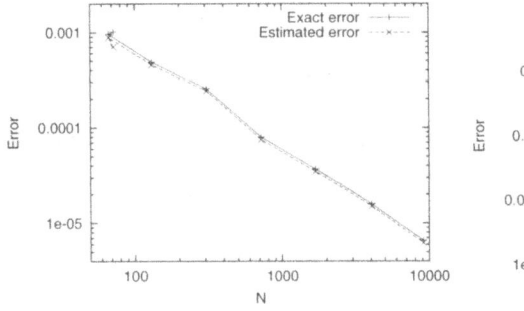 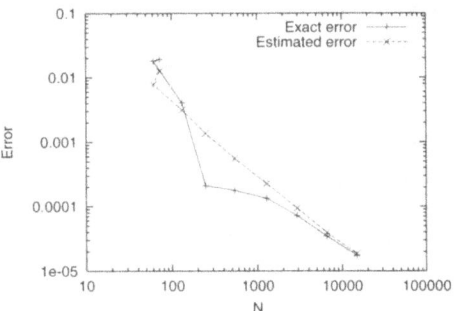

Figure A.5: *Quality of error estimates for the point value $u(a)$ (left) and the point derivative $\partial_1 u(a)$ (right).*

b) For the point value evaluation $J(u) = u(a)$, the convergence behavior of the error is shown in Figure A.6 for the following refinement criteria:

- Global refinement.

- Edge residuals: We drop the cell residual term from the error representation formula, and remove the weights involving the dual solution by assuming that there exists an a priori stability estimate for its second derivatives. We are then left with the following cell-wise term: $\eta_K = h_K^{1/2} \|[\partial_n u_h]\|_{\partial K}$.

- Weighted edge residuals: We replace the weights involving the dual solution by an a priori guess on each cell. Since for the point value, $\nabla^2 z$ decays as $r^{-2}$, where $r = |x - a|$, we take as indicator for refinement the expression $\eta_K = h_K^{1/2} \|[\partial_n u_h]\|_{\partial K} r_K^{-2}$.

- The full dual weighted error estimator $\eta_\omega$ .

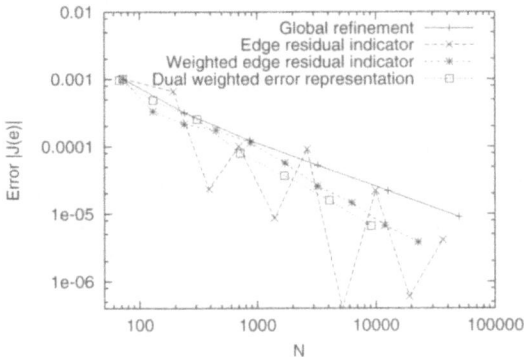

Figure A.6: *Comparison of various refinement criteria for the computation of the point value* $u(a)$.

As can be seen from the figure, a posteriori error indicators do a much better job than global refinement. Using the guessed weight is almost as good as the full weighted error estimator; the difference between the two can probably be attributed to the fact that the weight $r^{-2}$ does not 'see' the singularities of the dual solution at the reentrant corners of the domain, in contrast to the numerically computed approximation

The original edge residual indicator without weights displays a rather irregular behavior. The sign of the error is changing multiply, and the error grows at several refinement steps. The computed values are here, as opposed to the other criteria, not suitable for extrapolation.

# Chapter 4

**Exercise 4.1** Let $N_{old}$ be the old number of cells, and $X$ and $Y$ the fractions of cells to be refined and coarsened, respectively. Under (bisection-type) refinement, each cell is replaced by $2^d$ cells, while coarsening merges $2^d$ cells into one. All other cells remain as they are. The new number of cells is thus

$$N_{new} = (1 - X - Y + 2^d X + 2^{-d} Y) N_{old}.$$

a) In order to approximately double the number of cells, we have to choose fractions $0 \leq X, Y \leq 1, X + Y \leq 1$, such that $1 - X - Y + 2^d X + 2^{-d} Y \approx 2$. In 2-D, $X = \frac{1}{3} + \frac{1}{4} Y$ solves this.

b) To keep the number of cells roughly constant, $1 - X - Y + 2^d X + 2^{-d} Y \approx 2$ has to hold. In 2-D, the solution of this equation is $X = \frac{1}{4} Y$.

It should be noted that a cell can only then be coarsened if all its child cells are marked for coarsening. In practice, marking a fraction $Y$ of all cells for coarsening will therefore yield much less cells that will actually be unrefined. Thus, the computations above are only an indication for $N_{\mathrm{new}}$.

**Exercise 4.2** The proof of the cell-wise interpolation estimate

$$\|\nabla(u - I_h u)\|_K \leq c_I h_K \|\nabla^2 u\|_K$$

via the 'Bramble-Hilbert lemma' combines an abstract functional analytic result of the type of the Poincaré inequality with a scaling argument (see, e.g., Ciarlet [46]). Suppose for simplicity that the mesh is rectangular.

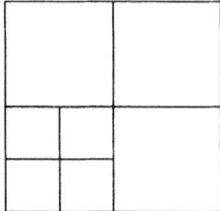

Figure A.7: *Rectangular cell patch with hanging nodes.*

The cell $K$ belongs to a patch $\tilde{K}$ of cells sharing with $K$ the hanging nodes (see Figure A.7), i.e., the interpolation $I_h u$ is well defined by $u_{|\tilde{K}}$. The cell patch $\tilde{K}$ of width $h_K$ is mapped by an affine mapping onto a reference domain $\tilde{K}_1$ of width one. Let $I_1$ denote the generic analogue of the interpolation operator $I_h$ on $\tilde{K}_1$. With this notation, we consider the functional

$$F(u) := \|\nabla(u - I_1 u)\|_{K_1},$$

which is well defined on the Sobolev space $H^2(\tilde{K}_1)$. It has the following three properties:

$$|F(u)| \leq c\,\|u\|_{H^2(\tilde{K}_1)}, \qquad \text{(boundedness)}$$
$$|F(u + v)| \leq |F(u)| + |F(v)|, \qquad \text{(subadditivity)}$$
$$F(q) = 0, \quad q \in P_1(\tilde{K}_1). \qquad \text{(orthogonality)}$$

The first one follows from the estimate

$$\|\nabla(u - I_h u)\|_{K_1} \leq \|\nabla u\|_{K_1} + c\|u\|_{L^\infty(\tilde{K}_1)}$$

which is a consequence of the definition of $I_h u$ and the equivalence of norms on finite dimensional vector spaces, together with the usual Sobolev inequality

$$\|u\|_{L^\infty(\tilde{K}_1)} \leq c\|u\|_{H^2(\tilde{K}_1)}.$$

The second one is obvious and the third one follows from the property $I_h u = u$ for $u \in P_1(K_1)$. Then, the abstract Bramble-Hilbert lemma implies that there holds

$$|F(u)| \leq \hat{c}\,\|\nabla^2 u\|_{\tilde{K}_1},$$

with a constant $\hat{c}$ depending only on the shape of the reference domain $\tilde{K}_1$. The asserted estimate then follows by a standard scaling argument from $\tilde{K}_1$ to $\tilde{K}$.

**Exercise 4.3 (practical exercise)** In FigureA.8, the error in the target functional $J(u) = \partial_1 u(a)$ is shown for both global refinement and refinement by the weighted error estimator. Obviously, the introduction of hanging nodes does not destroy the better convergence of the adapted meshes. In this case, one of the four cells adjacent to the evaluation point even has a hanging node on all levels of refinement. Yet, the results are better than as could be obtained on globally refined meshes with their regular mesh structure in the vicinity of the evaluation point. Removing this single hanging node produces better results but with a less smooth convergence behavior. It seems that also on locally refined meshes the *super-approximation* effect for the approximation of $\partial_1 u(a)$ is picked up, provided that the mesh is uniform in a small neighborhood of the point $a$.

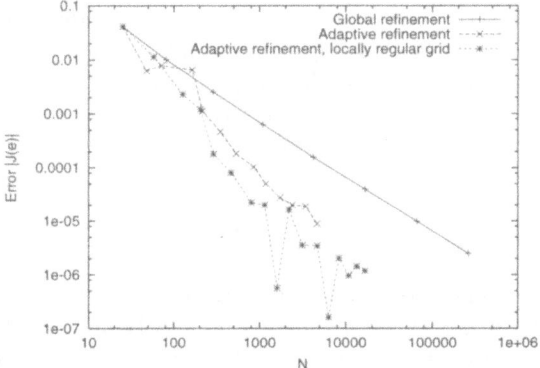

Figure A.8: *Error in the functional $\partial_1 u(a)$ for global refinement and refinement by the weighted error estimator.*

# Chapter 5

**Exercise 5.1** Choosing $\varepsilon = \mathrm{TOL}^{1/2}$ is actually possible here, since the difference between the point value and the integral mean over a domain which is symmetric to this point is quadratic in the diameter of this domain. A computation analogous to the one at the beginning of Chapter 5 shows that for general regularization parameters $\varepsilon$, the minimal mesh width satisfies $h_{\min}^2 = \varepsilon^{3/2} TOL$. Thus, for $\varepsilon = TOL$, we have $h_{\min} = TOL^{5/4}$, while for $\varepsilon = TOL^{1/2}$, there holds $h_{\min} = TOL^{7/8}$. Since each refinement step reduces the mesh width by a factor two (starting from an assumed coarsest cell of diameter $\approx 1$), the number of refinement steps necessary to reach $h_{\min}$ is $L = -\log_2 h_{\min}$. Thus, for the choices of $\varepsilon$, we get

$$L \approx \frac{5}{4} \frac{|\log TOL|}{|\log 2|}, \qquad \text{and} \qquad L \approx \frac{7}{8} \frac{|\log TOL|}{|\log 2|},$$

respectively. Therefore, the choice $\varepsilon = TOL^{1/2}$ reduces the necessary number of refinement steps by 30 per cent. Note that this choice of $\varepsilon$ is not possible if the point of evaluation is closer to the boundary than TOL.

**Exercise 5.2** Any reasonable choice for $\psi_h$ has to satisfy the following criteria:
(i) it must be from the discretization space $V_h$;
(ii) it should be locally constructible from $z$;
(iii) it should satisfy a local interpolation property such as

$$\|z - \psi_h\|_K \le c h_K^2 \|\nabla^2 z\|_K.$$

On the other hand, we would like to choose $\psi_h$ in such a way that the cell terms in the error representation are small compared to the edge terms. For simplicity, assume that the mesh consists solely of rectangles. Then $\Delta u_h = 0$, and the cell terms have the form $(f, z - \psi_h)_K$. Ideally, we would then like to choose $\psi$ such that it interpolates $z$ and that the mean value of $z - \psi_h$ is zero. In this case, the cell-wise mean value $\bar{f}$ of $f$ would be orthogonal to the weights, and we could estimate

$$(f, z - \bar{z})_K = (f - \bar{f}, z - \bar{z})_K \le \|f - \bar{f}\|_K \|z - \bar{z}\| \le c h_K^3 \|\nabla f\|_K \|\nabla^2 z\|_K,$$

making the term of higher order than the edge term. Unfortunately, this construction is not possible: if $\psi_h$ shall interpolate $z$ and be in $V_h$, then we do not have a degree of freedom left on each cell to satisfy the mean value property.

However, this construction is possible: let $\tilde{K}$ be a patch of four cells, then set $\psi_h(a_i) = z(a_i)$ on each of the eight vertices $a_i$ at the boundary of $\tilde{K}$, and determine the remaining degree of freedom at the center vertex such that $\int_{\tilde{K}} (z - \psi_h) \, dx = 0$. This choice of $\psi_h$ satisfies all the criteria listed above. The interpolation estimates and estimates for $f - \bar{f}$ now hold on $\tilde{K}$ instead if $K$, but as this is only an $\mathcal{O}(h)$ environment, this does not disturb us.

**Exercise 5.3 (practical exercise)** a) Figure A.9 shows the numerically computed approximation of the second derivatives for grids generated for the two functionals given in the exercise. Initially, the formula gives cell-wise constant values, but since this is a quantity that leads to poorly visible graphics, we generate a function where the value at each vertex is the mean value of the cell-wise quantities on the adjacent cells.

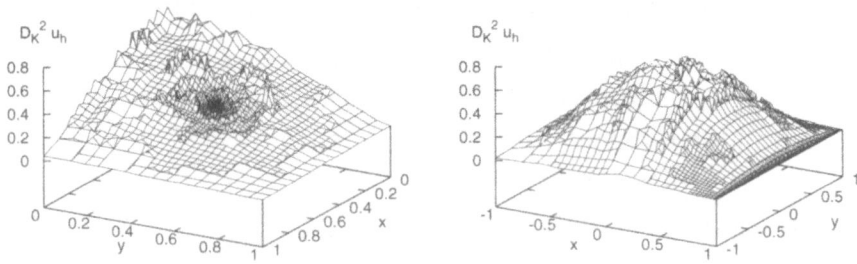

Figure A.9: $D_K^2 u_h$ on parts of optimal meshes generated by the dual weighted error estimator for $J(u) = \partial_1 u(a)$ (left) and $J(u) = \int_{-1}^{1} \partial_n u(1,y)dy$ (right).

It can be seen that the result is a relatively smooth function except for those cells with hanging nodes. At these, the computed value is larger than on patches of uniform grids, and the ratio with the value of the second derivative of the exact solution does not converge to one. However, computing on a sequence of successively finer grids, the maximal value of the approximation to the second derivatives remains bounded for both examples.

b) For the regular solution, the same holds as said above for the quotient of the computed approximation of the second derivative with the asymptotic exact second derivatives, shown in Figure A.10. Again, the values remain bounded, but are irregular at cells with hanging nodes.

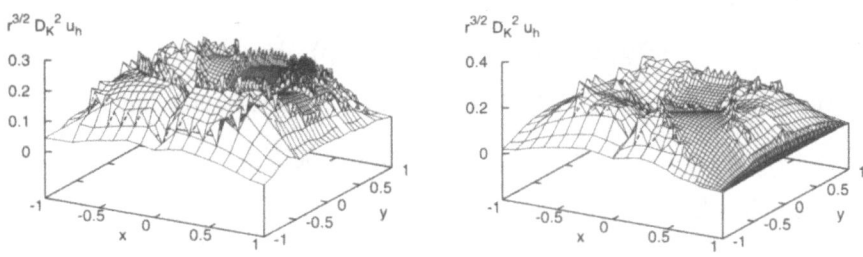

Figure A.10: $\tilde{D}_K^2 u_h$ on optimal meshes generated by the weighted error estimator for computing $J(u) = \partial_1 u(a)$ (left) and $J(u) = \int_{-1}^{1} \partial_n u(1,y)\,dy$ (right).

# Chapter 6

**Exercise 6.1** For the Burgers equation, the semilinear form and its derivatives are

$$A(u)(z) = \nu(u_x, z_x) + (uu_x, z) - (f, z)$$
$$A'(u)(w, z) = \nu(w_x, z_x) + (uw_x, z) + (wu_x, z)$$
$$A''(u)(\varphi, w, z) = (\varphi w_x, z) + (w\varphi_x, z)$$
$$A'''(u)(\psi, \varphi, w, z) = 0.$$

The second- and third-order remainder terms are thus

$$\mathcal{R}_h^{(2)} = \int_0^1 \left\{ A''(u_h + se)(e, e, z) - J''(u_h + se)(e, e) \right\} s \, ds$$

$$= \int_0^1 2(ee_x, z) \, s \, ds = (ee_x, z),$$

$$\mathcal{R}_h^{(3)} = \tfrac{1}{2} \int_0^1 \left\{ J'''(u_h + se)(e, e, e) - A'''(u_h + se)(e, e, e, z_h + se^*) \right.$$
$$\left. - 3A''(u_h + se)(e, e, e^*) \right\} s(s-1) \, ds$$

$$= -3 \int_0^1 (ee_x, e^*) \, s(s-1) \, ds = \tfrac{1}{2}(ee_x, e^*).$$

**Exercise 6.2** Use the output functional

$$J(\varphi) := \tfrac{1}{2} \|u - \varphi\|^2,$$

to obtain $J(u) - J(u_h) = -\tfrac{1}{2}\|e\|^2$. Its derivatives are

$$J'(v)(w) = (u - v, w), \quad J''(v)(w, \varphi) = -(w, \varphi).$$

The solution of the corresponding dual problem

$$(\nabla\varphi, \nabla z) = J'(u_h)(\varphi) = (e, \varphi) \quad \forall \varphi,$$

satisfies the a priori bound $\|\nabla^2 z\| \le c_S \|e\|$. This yields the error identity

$$-\tfrac{1}{2}\|e\|^2 = -\rho(u_h)(z - I_h z) + \mathcal{R}_h^{(2)},$$

with the second-order remainder term

$$\mathcal{R}_h^{(2)} = \int_0^1 \left\{ A''(u_h + se)(e, e, z) - J''(u_h + se)(e, e) \right\} s \, ds = \tfrac{1}{2}\|e\|^2.$$

Now proceed by bringing the remainder term to the left hand side, and using the standard argument to obtain

$$\|e\|^2 = \rho(u_h)(z - I_h z) \le c_I c_S \rho_{L^2}(u_h) \|e\|,$$

with the $L^2$-norm error estimator $\rho_{L^2}(u_h)$ as defined in Section 3.2. Note that we have actually made use of the remainder term in this example, and that it cannot be neglected here.

**Exercise 6.3 (practical exercise)** Figure A.11 shows the primal and dual solutions for the mean-value functional.

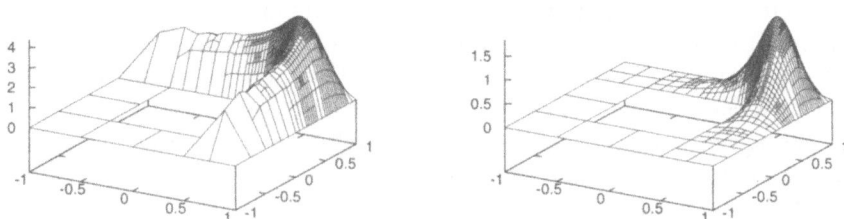

Figure A.11: *Primal and dual solutions for the mean-value functional.*

Figure A.12 shows the ratio of the nonlinearity indicator $\Delta\rho$ and the (second-order error estimate) $(\rho + \rho^*)/2$. In the left graph, this ratio is shown when primal and dual weights are approximated by a patch-wise higher-order interpolation. The result is a rather small value of the nonlinearity indicator, much less than 1 per cent even as we approach the critical eigenvalue $\alpha \approx 72.3$ (see also the solution to exercise 7.4).

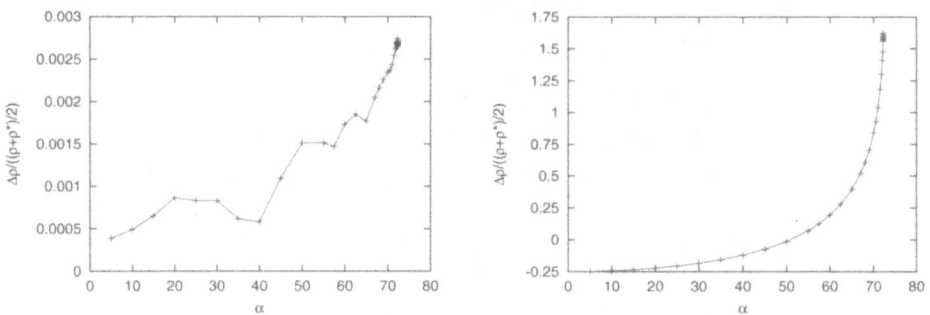

Figure A.12: *Ratio $\Delta\rho/((\rho + \rho^*)/2)$ between the nonlinearity term and the error estimate. Left: Approximation of primal and dual weights by patch-wise biquadratic interpolation. Right: Approximation of the dual weight by a higher-order Ritz projection. Note the difference in scale.*

However, as the right graph of the figure indicates, the above result is misleading: if the dual solution is approximated by a higher-order Ritz projection, then $\rho$ and $\rho^*$ are no more as close together, and $\Delta\rho$ becomes large, especially as we approach the critical eigenvalue. Close to the critical value, the difference between the two residuals becomes larger than the residuals themselves, making the error estimate quite unreliable.

The interpretation of this is that if both primal and dual weights are approximated using a higher-order interpolation, then the approximated residuals $\tilde{\rho}$ and $\tilde{\rho}^*$ are both not particularly close to the exact error, but the effects of approximation shift their values in the same direction, thus canceling out and pretending a small $\Delta\rho$. On the other hand, when replacing $z$ by its biquadratic Ritz projection, $\rho$ is computed to a high accuracy, while $\rho^*$ is still inaccurate because the weights in this term are only obtained by interpolation. Then, the approximation effect in $\rho^*$ is not canceled by that in $\rho$, and we obtain a large value for $\Delta\rho$.

In effect, the evaluation of $\Delta\rho$ does not tell us very much in this particular example, if we do not take into account the effects of numerically approximating the weights in the residuals.

# Chapter 7

**Exercise 7.1** We recall the defining equations

$$a(u,\psi) = \lambda\,(u,\psi) \quad \forall\psi \in V, \quad \|u\| = 1,$$
$$a(u_h,\psi_h) = \lambda_h\,(u_h,\psi_h) \quad \forall\psi_h \in V_h, \quad \|u_h\| = 1.$$

Consequently,

$$
\begin{aligned}
\lambda - \lambda_h &= \tfrac{1}{2}(\lambda-\lambda_h)\{\|u_h\|^2+\|u\|^2\} \\
&= \tfrac{1}{2}(\lambda-\lambda_h)\{\|u_h\|^2-2(u_h,u)+\|u\|^2\} + (\lambda-\lambda_h)(u_h,u) \\
&= \tfrac{1}{2}(\lambda-\lambda_h)\|u-u_h\|^2 + \lambda(u_h,u) - \lambda_h(u_h,u) \\
&= \tfrac{1}{2}(\lambda-\lambda_h)\|u-u_h\|^2 + a(u_h,u) - \lambda_h(u_h,u) \\
&= \tfrac{1}{2}(\lambda-\lambda_h)\|u-u_h\|^2 + \rho(u_h,\lambda_h)(u).
\end{aligned}
$$

Using Galerkin orthogonality $\rho(u_h,\lambda_h)(\psi_h) = 0$, $\psi_h \in V_h$, the asserted representation follows.

**Exercise 7.2** We recall the estimate (7.12) from Proposition 7.3,

$$|\lambda-\lambda_h| \leq \eta_\lambda^\omega := \sum_{K\in\mathbb{T}_h} \{\rho_K\omega_K^* + \rho_K^*\omega_K\},$$

$$\rho_K := \left( \|R_h\|_K^2 + h_K^{-1} \|r_h\|_{\partial K}^2 \right)^{1/2},$$

$$\rho_K^* := \left( \|R_h^*\|_K^2 + h_K^{-1} \|r_h^*\|_{\partial K}^2 \right)^{1/2}$$

$$\omega_K^* := \left( \|u^* - I_h u^*\|_K^2 + h_K \|u^* - I_h u^*\|_{\partial K}^2 \right)^{1/2},$$

$$\omega_K := \left( \|u - I_h u\|_K^2 + h_K \|u - I_h u\|_{\partial K}^2 \right)^{1/2}.$$

It remains to estimate the weights $\omega_K$ and $\omega_K^*$. By the usual estimates for the nodal interpolations $I_h u$ and $I_h u^*$, we have

$$\omega_K^* \leq c_I h_K^2 \|\nabla^2 u^*\|_K, \qquad \omega_K \leq c_I h_K^2 \|\nabla^2 u\|_K.$$

This gives us the intermediate result

$$|\lambda - \lambda_h| \leq c_I \left( \sum_{K \in \mathbb{T}_h} h_K^4 \rho_h^2 \right)^{1/2} \|\nabla^2 u^*\| + c_I \left( \sum_{K \in \mathbb{T}_h} h_K^4 \rho_h^{*2} \right)^{1/2} \|\nabla^2 u\|.$$

Since $u$ and $u^*$ are eigenfunctions of the operators $\mathcal{A} := -\Delta + b \cdot \nabla$ and $\mathcal{A}^* := -\Delta - b \cdot \nabla$, respectively, there holds

$$\|\nabla^2 u\| + \|\nabla^2 u^*\| \leq c_1(\mathcal{A}) \left\{ \|\mathcal{A}u\| + \|\mathcal{A}^* u^*\| \right\} = c_1(\mathcal{A}) |\lambda| \left\{ \|u\| + \|u^*\| \right\},$$

with some constant $c_1(\mathcal{A})$. Since $\|u^*\| \leq c_2(\mathcal{A})$, the asserted estimate follows.

**Exercise 7.3** Let $w(t)$ solve the (nonstationary) *perturbation equation*

$$\partial_t w - \nu \Delta w + \hat{u} \cdot \nabla w + w \cdot \nabla \hat{u} = 0,$$

corresponding to some initial perturbation $w_{|t=0} = w_0$. Taking the scalar product with $w$, and observing that

$$(\hat{u} \cdot \nabla w, w) = \tfrac{1}{2} (\hat{u}, \nabla w^2) = -\tfrac{1}{2} (\nabla \cdot \hat{u}, w^2)$$

yields

$$\tfrac{1}{2} d_t \|w\|^2 + \nu \|\nabla w\|^2 + (\{\nabla \hat{u}^T - \tfrac{1}{2} \nabla \cdot \hat{u} I\} w, w) = 0, \quad t \geq 0.$$

If now

$$\nu \|\nabla w\|^2 + (\{\nabla \hat{u}^T - \tfrac{1}{2} \nabla \cdot \hat{u} I\} w, w) \geq 0,$$

then integration with respect to $t$ implies

$$\tfrac{1}{2} \|w(t)\|^2 - \tfrac{1}{2} \|w_0\|^2 \leq 0,$$

i.e., the $L^2$ norm of the perturbation $w(t)$ stays bounded. Hence, a sufficient criterion for the $L^2$ stability of the base solution $\hat{u}$ is that all eigenvalues $\lambda$ of the following *symmetric* eigenvalue problem in $V$ are nonnegative:

$$-\nu \Delta w + \tfrac{1}{2} \{\nabla \hat{u} + \nabla \hat{u}^T - \nabla \cdot \hat{u} I\} w = \lambda w.$$

**Exercise 7.4 (practical exercise)** In the left part of Figure A.13, the stability eigenvalue $\lambda_{\min}$ is shown. As the forcing strength $\alpha$ approaches the critical value $\alpha_{\mathrm{crit}} \approx 72.3$, the solution $\hat{u}$ of the base problem $-\Delta\hat{u} - \hat{u}^3 = \alpha$ grows until the smallest eigenvalue of the operator $-\Delta - 3\hat{u}^2$ becomes zero.

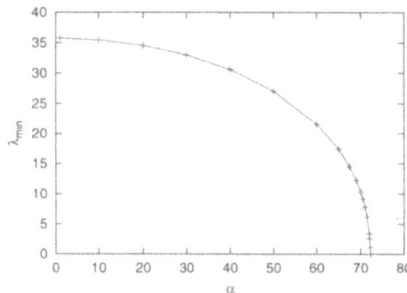 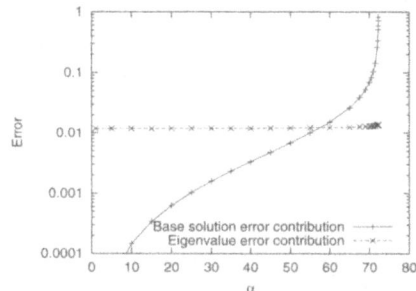

Figure A.13: *Numerically computed stability eigenvalues $\lambda_{\min}$ as function of the forcing strength $\alpha$ (left). Base solution error contribution $\hat{\eta}_\lambda = \frac{1}{2}\hat{\rho} + \frac{1}{2}\hat{\rho}^*$ and eigenvalue approximation error contribution $\eta_\lambda = \frac{1}{2}\rho + \frac{1}{2}\rho^*$ (right).*

For the computation of this stability eigenvalue, we want to use the error representation formula for $\lambda - \lambda_h$, consisting of the terms $\hat{\eta}_\lambda = \frac{1}{2}\hat{\rho} + \frac{1}{2}\hat{\rho}^*$ for the accuracy of the base solution, and $\eta_\lambda = \frac{1}{2}\rho + \frac{1}{2}\rho^*$ for the accuracy of the eigenvalue approximation. Note that $\rho = \rho^*$ for the symmetric problem under consideration here. In the right part of Figure A.13, the sizes of $\hat{\eta}_\lambda$ and $\eta_\lambda$ are shown for a fixed, uniformly refined grid with roughly 50,000 cells.

As can be seen from the comparison of the two components of the error estimator, the error due to the approximation of the eigenvalue grows only very moderately as $\alpha \to \alpha_{\mathrm{crit}}$. On the other hand, the error contribution due to the approximation of the base solution becomes very large, eventually to the same order of magnitude as the eigenvalue itself on this particular grid when near to the critical value of $\alpha$. This then indicates that the present grid is entirely unsuitable for the computation of a stability eigenvalue. It should be noted that for all values of $\alpha$, the estimated error is indeed relatively close to the true error.

# Chapter 8

**Exercise 8.1** The Lagrangian for this problem has the form

$$L(u, q, \lambda) = J(u) - A(u)(\lambda) = \tfrac{1}{2}\|u - \bar{u}\|^2 - (\nabla u, \nabla \lambda) - (qu, \lambda) - (f, \lambda).$$

The error estimator for $J(u, q) - J(u_h, q_h) = J(u) - J(u_h)$ reads

$$J(u) - J(u_h) = \tfrac{1}{2}L'(u_h, q_h, \lambda_h)(u - \varphi_h, q - \chi_h, \lambda - \psi_h) + \mathcal{R}_h,$$

for arbitrary test functions $\varphi_h, \chi_h, \psi_h$, and a remainder term $\mathcal{R}_h$ cubic in the errors. Since only one term of the Lagrangian is cubic, the remainder is a multiple of $(e^q e^u, e^\lambda)$. Also, since $q \in \mathbb{R}$, the corresponding test function $\chi_h$ is a scalar as well, and we can make the residual term $L'_q(u_h, q_h, \lambda_h)(q - \chi_h)$ to zero by choosing $\chi_h = q$. For the rest of the estimator, we introduce residual and weight terms

$$\rho_K^u = \left(\|f + \Delta u_h - q_h u_h\|_K^2 + \tfrac{1}{4}h_K^{-1}\|[\partial_n u_h]\|_{\partial K}^2\right)^{1/2},$$

$$\rho_K^\lambda = \left(\| -(u_h - \bar{u}) + \Delta\lambda_h - q_h\lambda_h\|_K^2 + \tfrac{1}{4}h_K^{-1}\|[\partial_n \lambda_h]\|_{\partial K}^2\right)^{1/2},$$

$$\omega_K^u = \left(\|u - I_h u\|_K^2 + h_K\|u - I_h u\|_{\partial K}^2\right)^{1/2},$$

$$\omega_K^\lambda = \left(\|\lambda - I_h\lambda\|_K^2 + h_K\|\lambda - I_h\lambda\|_{\partial K}^2\right)^{1/2},$$

to obtain

$$J(u) - J(u_h) = \sum_{K \in \mathbb{T}_h} \{\rho_K^u \omega_K^\lambda + \rho_K^\lambda \omega_K^u\}.$$

**Exercise 8.2** The Lagrangian for this problem reads

$$L(u, \lambda, z) = J(u) - A(u)(z) = \tfrac{1}{2}(u - 1, 1)^2 - (\nabla u, \nabla z) + \lambda(u, z).$$

From this, we obtain the optimality conditions determining a solution to the constrained optimization problem:

$$L'_u(u, \lambda, z)(\varphi) = (u - 1, 1)(\varphi, 1) - (\nabla\varphi, \nabla z) + \lambda(\varphi, z) = 0,$$
$$L'_\lambda(u, \lambda, z)(\mu) = (u, z) = 0,$$
$$L'_z(u, \lambda, z)(\psi) = -(\nabla u, \nabla\psi) + \lambda(u, \psi) = 0,$$

for all test functions $\{\varphi, \mu, \psi\}$. In strong form, the first and last equations read

$$-\Delta u - \lambda u = 0, \quad -\Delta z - \lambda z = (u - 1, 1).$$

The first determines the eigenvalue $\lambda$ and the eigenfunction $u$ up to a multiplicative constant. Since the second optimality condition above requires that $z$ has no component in direction of $u$, this removes the kernel of the operator $-\Delta-\lambda$, using the assumption of simplicity of this eigenvalue. Furthermore, the adjoint equation has a solution only if the right hand side (which is a constant) is in the range of the operator, $\mathcal{R}(-\Delta-\lambda) = \mathcal{N}(-\Delta-\lambda)^{\perp} = \{u\}^{\perp}$; this condition is only satisfied if either the constant right hand side is zero, or if $u$ has mean value zero. Since it is known that the lowest eigenfunction of the Laplacian does not change its sign, the second case cannot occur. In the first case, $u$ is normalized, and the solution of the adjoint equation is necessarily $z \equiv 0$. Since in the error representation formula all terms contain $z$ multiplicatively, the error is zero, for every grid chosen.

The reason for this surprising behavior is that we have set out to control the error in the functional $J(u) = \frac{1}{2}\{(u,1)-1\}^2$. For the exact minimizer, $J(u) = 0$. On the other hand, we can find discrete eigenvalue/eigenvector pairs on *every* grid, and can normalize the eigenvector to $(u_h,1) = 1$. So $J(u_h)$ can be made equal to zero as well on each grid, making $J(u)-J(u_h) = 0$. Thus, the error estimator only tells us that we can make the error in this quantity to zero on every grid.

**Exercise 8.3 (practical exercise)** In Figure A.14, two grids generated by an energy-norm error estimator and the weighted error estimator are compared. It is obvious that the latter takes into account that the solution needs to be represented accurately at the observation boundary $\Gamma_O$ at the right, while the energy-norm error estimator only sees the structure of the primal solution, in particular near the control boundary $\Gamma_C$ at the bottom, and at the corners next to it. Primal and adjoint solutions obtained on the second grid are shown in Figure A.15.

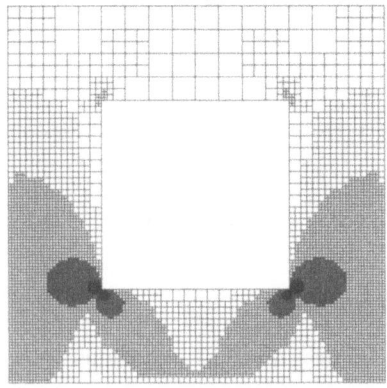 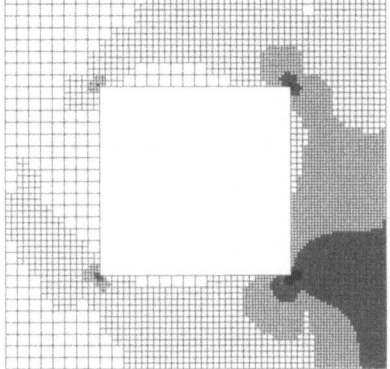

Figure A.14: *Grids generated by an energy-norm error estimator for the primal solution (left) and by the weighted error estimator (right), with approximately 9,000 cells each. Control boundary at the bottom, observation at the right.*

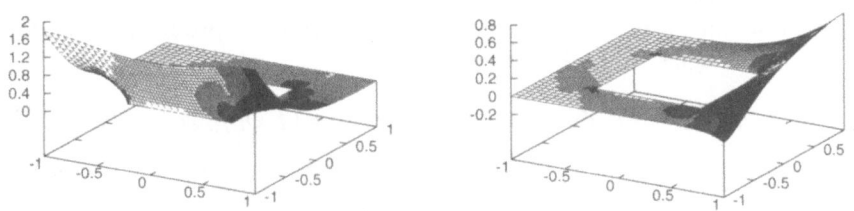

Figure A.15: *Primal (left) and adjoint solution (right) computed on the right grid of Figure A.14.*

Finally, we check the quality of the generated meshes. In the left part of Figure A.16, the error in the optimal control parameter, $|q - q_h|$, is shown. Obviously, the grids generated by the weighted error estimator are not as efficient in reducing this error as the energy-norm error estimator. However, this should not surprise us, since the weighted estimator used the natural output functional, $J(u) = \frac{1}{2}\|u - \bar{u}\|^2_{\Gamma_O}$, rather than a functional involving the control parameter $q$. On the other hand, the right part of Figure A.16 shows clearly that the meshes generated by the DWR method are superior in reducing the error in the functional for which it was made. Making the DWR method superior also for the error in $q$ would require to solve a dual problem with a different output functional.

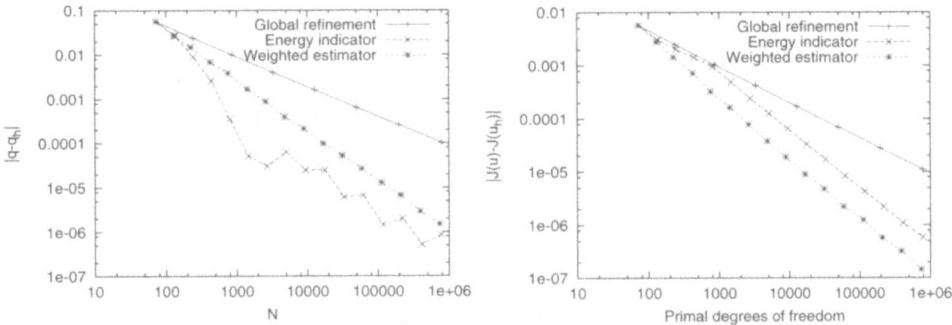

Figure A.16: *Error in the control parameter q (left) and in the misfit functional J (right) for different mesh refinement criteria.*

# Chapter 9

**Exercise 9.1** Since trial functions are piecewise linear and continuous in time, while test functions are piecewise constant, the discretized problem decouples into a system of equations for each time-step:

$$(u_h^n - u_h^{n-1}, \psi_h)_\Omega + \tfrac{1}{2} k_n (\nabla(u_h^n + u_h^{n-1}), \nabla \psi_h)_\Omega = (f, \psi_h)_{\Omega \times I_n} \qquad \forall \psi_h \in V_h^n,$$
$$(u_h^0, \psi_h)_\Omega = (u^0, \psi_h)_\Omega \qquad \forall \psi_h \in V_h^n,$$

Thus, the error estimator has the form

$$J(e) = \sum_{n=1}^N \sum_{K \in T_h^n} (f - \partial_t u_h + \Delta u_h, z - z_h)_{K \times I_n} + \tfrac{1}{2}([\partial_n u_h], z - z_h)_{\partial K \times I_n}$$
$$+ (u_h^0 - u^0, (z - z_h)_{|t=0})_\Omega .$$

For the end-time error, we take the scalar products of residuals and weights apart and use a priori estimates for $z - z_h$ to eliminate the dual solution. Since the test space from which $z_h$ is taken is the same as for the dG(0) method, the same steps as in Section 9.2 can be used to accomplish this.

**Exercise 9.2** The given formulation differs from the one used in Section 9.3 only in the way the velocity equation $\partial_t w = v$ is enforced: instead of $L^2$-scalar products with test functions, we use the Dirichlet form. Thus, the only difference to the estimator given in the main text is the form of the residual of the first equation, which now reads

$$R_h^w = -\Delta(\partial_t w_h - v_h),$$

and the addition of another term $(r_h^w, z^w - \varphi_h^w)_{\partial K \times I_n}$ with jump residuals

$$r_h^w {}_{|\Gamma \times I_n} = \begin{cases} -\tfrac{1}{2}[\partial_n(\partial_t w_h - v_h)] & \text{if } \Gamma \not\subset \partial\Omega, \\ 0 & \text{otherwise.} \end{cases}$$

**Exercise 9.3 (practical exercise)** Since the (negative) Laplace operator possesses a complete orthonormal basis $\{v_i\}_{i=1}^\infty$ of eigenvectors with corresponding (positive) eigenvalues $\lambda_i$, the solution of the heat equation is given by

$$u = \sum_{i=1}^\infty \lambda_i^{-1} (1, w_i)_\Omega \left(1 - e^{-\lambda_i t}\right) w_i.$$

As the lowest eigenfunction of the Laplace operator, $w_1$, is the only one that does not change sign, and since $\lambda_1$ is the smallest eigenvalue, the respective coefficient is the largest for this given right-hand side. Thus, the dynamics of the solution can be well approximated by only considering this particular mode, and the choice of a fixed grid is justified if we are not interested in the initial behavior.

Consequently, if $h$ is small enough (e.g., $h^2 = \mathcal{O}(k)$), we have for the end-time error due to equation (9.18):

$$\|e^{N-}\|^2 \leq c \sum_{n=1}^{N} \sigma_n^{-1} \tau_n^{-1} k_n^2 (\rho_k^n)^2,$$

with the total time residual $(\rho_k^n)^2 := \sum_{K \in \mathbb{T}_h} (\rho_{K,k}^n)^2$. Using the definition of $\sigma$, neglecting the effect of $\tau$, and approximating the residual by the first mode, we obtain

$$\|e^{N-}\|^2 \leq c \sum_{n=1}^{N} e^{-\lambda_1 T} e^{\lambda_1 t_n} k_n^3 e^{-2\lambda_1 t_n} = c \sum_{n=1}^{N} e^{-\lambda_1 T} e^{-\lambda_1 t_n} k_n^3.$$

An error-balancing strategy would thus choose

$$k_n \propto \frac{TOL^2}{N} e^{\lambda_1 T/3} e^{\lambda_1 t_n/3},$$

resulting in a complexity of $N \propto TOL^{-1} e^{-\lambda_1 T/2}$. The difference to the example shown in Section 9.2, where $k_n$ decreases towards the end time, stems from the fact that there $\|\partial_t u(t)\| = \mathcal{O}(1)$.

# Chapter 10

**Exercise 10.1** Starting at the present iterate $u_h^k$, the next Newton update direction $\delta u_h^k$ is computed using the equation

$$a'(u_h^k)(\delta u_h^k, \psi_h) = -a(u_h^k)(\psi_h) \qquad \forall \psi_h \in V_h.$$

Since the semilinear form $a(\cdot)(\cdot)$ is not differentiable, we approximate the above equation by

$$\tilde{a}'(u_h^k)(\delta u_h^k, \psi_h) = -a(u_h^k)(\psi_h) \qquad \forall \psi_h \in V_h,$$

with $\tilde{a}'(u_h^k)(\delta u_h^k, \psi_h) = (\tilde{C}'(\varepsilon(u_h^k))\delta u_h^k, \varepsilon(\psi_h))$, and $\tilde{C}'$ as defined in (10.12).

The solvability of the Newton update equation can be shown in the same way as that of the linear elastic equations. For this, note that there the bilinear form reads $(2\mu\varepsilon(u)^D + \kappa \mathrm{tr}(\varepsilon(u))I, \varepsilon(\psi))$, where here, using the definition of $\tilde{C}'$, we only have to replace the coefficient $\mu$ by the spatially varying coefficient

$$\tilde{\mu} = \begin{cases} \mu I & \text{where } |\tau^D| = |\varepsilon(u_h^k)^D| \leq \sigma_0, \\ \frac{1}{2}\frac{\sigma_0}{|\tau^D|}\left\{\delta_{ij}\delta_{kl} - \frac{\tau_{ij}^D \tau_{kl}^D}{|\tau^D|^2}\right\} & \text{where } |\tau^D| = |\varepsilon(u_h^k)^D| > \sigma_0. \end{cases}$$

It thus remains to show that the fourth order tensor in the second line is positive definite, i.e.

$$\varepsilon_{ij}^{D}\Big\{\delta_{ij}\delta_{kl}-\frac{\tau_{ij}^{D}\tau_{kl}^{D}}{|\tau^{D}|^{2}}\Big\}\varepsilon_{kl}^{D}\geq 0,$$

for all $\varepsilon^{D}$. This, however, is obvious.

**Exercise 10.2** We start from the 'continuous' inf-sup-stability estimate applied for $p_h \in Q_h \subset Q$:

$$\sup_{u\in V}\frac{(p_h,\nabla\!\cdot\!u)}{\|\varepsilon(u)\|}\geq\gamma\|p_h\|,\quad p_h\in Q_h\,.$$

With an $H^1$-stable local interpolant $\tilde{I}_h u \in V_h$, there holds

$$\frac{(p_h,\nabla\!\cdot\!u)}{\|\varepsilon(u)\|}=\frac{(p_h,\nabla\!\cdot\!(u-\tilde{I}_h u))}{\|\varepsilon(u)\|}+\frac{(p_h,\nabla\!\cdot\!\tilde{I}_h u)}{\|\varepsilon(u)\|}$$

$$=-\frac{(\nabla p_h,u-\tilde{I}_h u)}{\|\varepsilon(u)\|}+\frac{(p_h,\nabla\!\cdot\!\tilde{I}_h u)}{\|\varepsilon(\tilde{I}_h u)\|}\frac{\|\varepsilon(\tilde{I}_h u)\|}{\|\varepsilon(u)\|}\,.$$

Noting the estimates

$$-\frac{(\nabla p_h,u-\tilde{I}_h u)}{\|\varepsilon(u)\|}\geq-c\sum_{K\in\mathbb{T}_h}\frac{\|\nabla p_h\|_K\|u-\tilde{I}_h u\|_K}{\|\nabla u\|}\geq-c\Big(\sum_{K\in\mathbb{T}_h}h_K^2\|\nabla p_h\|_K^2\Big)^{1/2},$$

and, by Korn inequality,

$$\frac{\|\varepsilon(\tilde{I}_h u)\|}{\|\varepsilon(u)\|}\leq c\frac{\|\nabla(\tilde{I}_h u)\|}{\|\nabla u\|}\leq c,$$

we obtain the inf-sup stability estimate

$$\sup_{u_h\in V_h}\frac{(p_h,\nabla\!\cdot\!u_h)}{\|\varepsilon(u_h)\|}+\Big(\sum_{K\in\mathbb{T}_h}\delta_K\|\nabla p_h\|_K^2\Big)^{1/2}\geq\gamma\,\|p_h\|,\quad p_h\in Q_h\,,$$

with $\delta_K=\alpha h_K^2$ and $\alpha>0$ sufficiently large.

# Chapter 11

**Exercise 11.1** The surface-oriented form of the drag coefficient reads

$$c_{\text{drag}} = \frac{2}{\bar{U}^2 D} \int_S n^T (2\nu\tau - pI) e^{(1)} \, ds,$$

where $\tau = \frac{1}{2}(\nabla v + \nabla v^T)$. The divergence theorem states that $\int_\Omega \nabla \cdot f \, dx = \int_{\partial\Omega} n \cdot f \, do$. Therefore, we identify $f$ with $(2\nu\tau - pI)\bar{e}^{(1)}$, where we extend $e^{(1)}$ away from $S$ into the domain by $\bar{e}^{(1)}$. Thus,

$$c_{\text{drag}} = \frac{2}{\bar{U}^2 D} \int_{\partial\Omega} \nabla \cdot [(2\nu\tau - pI)\bar{e}^{(1)}] \, dx - \frac{2}{\bar{U}^2 D} \int_{\partial\Omega \setminus S} n^T (2\nu\tau - pI)\bar{e}^{(1)} \, ds.$$

Up to now, $\bar{e}^{(1)}$ is an arbitrary extension of $e^{(1)}$. In order to eliminate the second integral, we choose $\bar{e}^{(1)}$ such that it vanishes on $\partial\Omega \setminus S$. As long as we compute the drag coefficient of an object with finite distance to $\partial\Omega \setminus S$, $\bar{e}^{(1)}$ can be chosen to have finite gradient, thus avoiding problems with the definition of the domain integral.

Still, we do not want to compute $\nabla \cdot (2\nu\tau - pI)$ numerically since this involves second derivatives of the numerically computed solution. We therefore use that $\nabla\tau = \frac{1}{2}(\Delta v + \nabla\nabla \cdot v)$, and $\nabla \cdot (pI) = \nabla p$, and the state equation to rewrite $\nabla \cdot (2\nu\tau - pI) = -f + v \cdot \nabla v$, and thus obtain

$$c_{\text{drag}} = \frac{2}{\bar{U}^2 D} \int_{\partial\Omega} (v \cdot \nabla v - f) \cdot \bar{e}^{(1)} + (2\nu\tau - pI) : \nabla\bar{e}^{(1)} \, dx.$$

The computation for the lift coefficient follows the same procedure.

**Exercise 11.2** For the given problem, the primal variational formulation uses the semilinear form

$$a(u)(\psi) = \nu(\nabla v, \nabla\psi^v) + (v \cdot \nabla v, \psi^v) - (p, \nabla \cdot \psi^v) - (v, \nabla\psi^p) - (f, \psi^v).$$

The dual problem is thus characterized by $a'(u)(\varphi, z) = J(\varphi)$, which reads in weak form:

$$\nu(\nabla\varphi^v, \nabla z^v) + (v \cdot \nabla\varphi^v, z^v) + (\varphi^v \cdot \nabla v, z^v) - (\varphi^p, \nabla \cdot z^v) - (\varphi^v, \nabla z^p) = J(\varphi).$$

In strong form, this leads to the equations

$$-\nu\Delta z^v - (\nabla \cdot v)z^v - v \cdot \nabla z^v + (\nabla v)z^v + \nabla z^p = j,$$
$$\nabla \cdot z^v = 0.$$

Note that for the continuous dual problem, the second term in the first equation is actually zero, due to the incompressibility of $v$, but that it exists in the equation defining the discrete dual solution $z_h$. With the above, the residuals are

$$R^*_{h|K} = j + \nu\Delta z^v_h + (\nabla\cdot v_h)z^v_h + v_h\cdot\nabla z^v_h - (\nabla v_h)z^v_h + \nabla z^p_h$$

$$r^*_{h|\Gamma} = \begin{cases} \frac{1}{2}[\nu n\partial_n z^v_h + (v_h\cdot n)z^v_h - nz^p_h], & \text{if } \Gamma \not\subset \partial\Omega, \\ -\nu n\partial_n z^v_h - (v_h\cdot n)z^v_h + nz^p_h, & \text{if } \Gamma \subset \Gamma_{\text{out}}, \\ 0, & \text{otherwise.} \end{cases}$$

# Bibliography

[1] M. Ainsworth and J. T. Oden. A unified approach to a posteriori error estimation using element residual methods. *Numer. Math.*, 65:23–50, 1993.

[2] M. Ainsworth and J. T. Oden. A posteriori error estimation in finite element analysis. *Comput. Methods Appl. Mech. Eng.*, 142:1–88, 1997.

[3] M. Ainsworth and J. T. Oden. A posteriori error estimators for the Stokes and Oseen equations. *SIAM J. Numer. Anal.*, 34:228–245, 1997.

[4] J. P. Aubin. Behaviour of the error of the approximate solutions of boundary value problems for linear elliptic operators by Galerkin's and finite difference methods. *Ann. Scuola Morm. Sup. Pisa*, 21:599–637, 1967.

[5] I. Babuška and A. D. Miller. The post-processing approach in the finite element method, I: calculations of displacements, stresses and other higher derivatives of the displacements. *Int. J. Numer. Meth. Eng.*, 20:1085–1109, 1984.

[6] I. Babuška and A. D. Miller. The post-processing approach in the finite element method, II: the calculation of stress intensity factors. *Int. J. Numer. Meth. Eng.*, 20:1111–1129, 1984.

[7] I. Babuška and A. D. Miller. The post-processing approach in the finite element method, III: a posteriori error estimation and adaptive mesh selection. *Int. J. Numer. Meth. Eng.*, 20:2311–2324, 1984.

[8] I. Babuška and A. D. Miller. A feedback finite element method with a posteriori error estimation. *Comput. Methods Appl. Mech. Eng.*, 61:1–40, 1987.

[9] I. Babuška and W. C. Rheinboldt. Error estimates for adaptive finite element computations. *SIAM J. Numer. Anal.*, 15:736–754, 1978.

[10] I. Babuška and W. C. Rheinboldt. A posteriori error estimates for the finite element method. *Int. J. Numer. Meth. Eng.*, 12:1597–1615, 1978.

[11] I. Babuška and C. Schwab. A posteriori error estimation for hierarchic models of elliptic boundary value problems on thin domains. *SIAM J. Numer. Anal.*, 33:221–246, 1996.

[12] I. Babuška and T. Strouboulis. *The Finite Element Method and its Reliability.* Clarendon Press, Oxford, 2001.

[13] E. Backes. Gewichtete a posteriori Fehleranalyse bei der adaptiven Finite-Elemente-Methode: Ein Vergleich zwischen Residuen- und Bank-Weiser-Schätzer. Diploma thesis, Institute of Applied Mathematics, University of Heidelberg, 1997.

[14] W. Bangerth. Finite Element Approximation of the Acoustic Wave Equation: Error Control and Mesh Adaptation. Diploma thesis, Institute of Applied Mathematics, University of Heidelberg, 1998.

[15] W. Bangerth. Adaptive Finite Element Methods for the Identification of Distributed Parameters in Partial Differential Equations. Dissertation, Institute of Applied Mathematics, University of Heidelberg, 2002.

[16] W. Bangerth and R. Rannacher. Finite element approximation of the acoustic wave equation: Error control and mesh adaptation. *East–West J. Numer. Math.*, 7:263–282, 1999.

[17] R. E. Bank and A. Weiser. Some a posteriori error estimators for elliptic partial differential equations. *Math. Comp.*, 44:283–301, 1985.

[18] R. Becker. An Adaptive Finite Element Method for the Incompressible Navier–Stokes Equations on Time-Dependent Domains. Dissertation, Institute of Applied Mathematics, University of Heidelberg, 1995.

[19] R. Becker. An adaptive finite element method for the Stokes equations including control of the iteration error. *ENUMATH'97* (H. G. Bock *et al.*, eds), pp. 609–620, World Scientific, Singapore, 1998.

[20] R. Becker. An optimal-control approach to a posteriori error estimation for finite element discretizations of the Navier-Stokes equations. *East-West J. Numer. Math.*, 9:257–274, 2000.

[21] R. Becker. Mesh adaptation for stationary flow control. *J. Math. Fluid Mech.*, 3:317–341, 2001.

[22] R. Becker. Adaptive Finite Elements for Optimal Control Problems. Habilitation thesis, University of Heidelberg, 2001.

[23] R. Becker and M. Braack. Multigrid techniques for finite elements on locally refined meshes. *Numer. Linear Algebra Appl.*, 7:363–379, 2000.

[24] R. Becker and M. Braack. Solution of a stationary benchmark problem for natural convection with large temperature difference. *Int. J. Therm. Sci.*, 41:428–439, 2002.

[25] R. Becker, M. Braack, and R. Rannacher. Numerical simulation of laminar flames at low Mach number by adaptive finite elements. *Combust. Theory Modelling*, 3:503–534, 1999.

[26] R. Becker, M. Braack, R. Rannacher, and C. Waguet. Fast and reliable solution of the Navier–Stokes equations including chemistry. *Comput. Visual. Sci.*, 2:107–122, 1999.

[27] R. Becker, C. Johnson, and R. Rannacher. Adaptive error control for multigrid finite element methods. *Computing*, 55:271–288, 1995.

[28] R. Becker, H. Kapp, and R. Rannacher. Adaptive finite element methods for optimal control of partial differential equations: Basic concepts. *SIAM J. Control Optim.*, 39, 113–132, 2000.

[29] R. Becker and R. Rannacher. Weighted a posteriori error control in FE methods. Lecture at ENUMATH-95, Paris, Sept. 18-22, 1995, Preprint 96-01, SFB 359, University of Heidelberg, Proc. *ENUMATH'97* (H. G. Bock *et al.*, eds), pp. 621-637, World Scientific, Singapore, 1998.

[30] R. Becker and R. Rannacher. A feed-back approach to error control in finite element methods: Basic analysis and examples. *East-West J. Numer. Math.*, 4:237–264, 1996.

[31] R. Becker and R. Rannacher. An optimal control approach to error estimation and mesh adaptation in finite element methods. *Acta Numerica 2000* (A. Iserles, ed.), pp. 1-101, Cambridge University Press, 2001.

[32] R. Becker and B. Vexler. Adaptive finite element methods for parameter identification problems. Preprint 2002-20 (SFB 359), Universität Heidelberg, July 2002, *Numer. Math.*, submitted, 2002

[33] C. Bernardi, O. Bonnon, C. Langouët, and B. Métivet. Residual error indicators for linear problems: Extension to the Navier–Stokes equations. In Proc. *9th Int. Conf. Finite Elements in Fluids*, 1995.

[34] H. Blum, Q. Lin, and R. Rannacher. Asymptotic error expansion and Richardson extrapolation for linear finite elements. *Numer. Math.*, 49:11–37, 1986.

[35] H. Blum and F.-T. Suttmeier. An adaptive finite element discretization for a simplified Signorini problem. *Calcolo*, 37:65–77, 1999.

[36] H. Blum and F.-T. Suttmeier. Weighted error estimates for finite element solutions of variational inequalities. *Computing*, 65:119–134, 2000.

[37] K. Böttcher. Adaptive Schrittweitenkontrolle beim unstetigen Galerkin-Verfahren für gewöhnliche Differentialgleichungen. Diploma thesis, Institute of Applied Mathematics, University of Heidelberg, 1996.

[38] K. Böttcher and R. Rannacher. Adaptive error control in solving ordinary differential equations by the discontinuous Galerkin method. Technical Report Preprint 96-53, SFB 359, Universität Heidelberg, 1996.

[39] M. Braack. An Adaptive Finite Element Method for Reactive Flow Problems. Dissertation, Institute of Applied Mathematics, University of Heidelberg, 1998.

[40] M. Braack, R. Becker, and R. Rannacher. The dual-weighted residual method applied to three-dimensional flow problems. Computers & Fluids, Proc. 3. Int. Conf. on Appl. Math. for Industrial Flow Problems (AMIF), Lisbon, April 17-20, 2002, to appear.

[41] M. Braack and A. Ern. A posteriori control of modeling errors and discretization errors. Preprint 2002-13 (SFB 359), Universität Heidelberg, June 2002, SIAM J. Multiscale Modeling and Simulation, submitted, 2002.

[42] M. Braack and R. Rannacher. Adaptive finite element methods for low-Mach-number flows with chemical reactions. In H. Deconinck, editor, *30th Computational Fluid Dynamics, Lecture Series*, Vol. 1999-03, von Karman Institute for Fluid Dynamics, 1999.

[43] S. Brenner and R. L. Scott. *The Mathematical Theory of Finite Element Methods*. Springer, Berlin Heidelberg New York, 1994.

[44] G. F. Carey and J. T. Oden. *Finite Elements, Computational Aspects*, Vol. III. Prentice-Hall, 1984.

[45] C. Carstensen and R. Verfürth. Edge residuals dominate a posteriori error estimators for low-order finite element methods. *SIAM J. Numer. Anal.*, 36:1571–1587, 1999.

[46] P. G. Ciarlet. *Finite Element Methods for Elliptic Problems*. North-Holland, Amsterdam, 1978.

[47] R. Courant. Variational methods for the solution of problems of equilibrium and vibrations. *Bull. Amer. Math. Soc.*, 49:1–23, 1943.

[48] W. Dörfler and M. Rumpf. An adaptive strategy for elliptic problems including a posteriori controlled boundary approximation. *Math. Comp.*, 224:1361–1382, 1998.

[49] T. Dunne. Adaptive dual-gemischte Finite-Elemente-Verfahren. Diploma thesis, Institute of Applied Mathematics, University of Heidelberg, 2001.

[50] K. Eriksson, D. Estep, P. Hansbo, and C. Johnson. Introduction to adaptive methods for differential equations. *Acta Numerica 1995* (A. Iserles, ed.), pp. 105–158, Cambridge University Press, 1995.

[51] K. Eriksson and C. Johnson. An adaptive finite element method for linear elliptic problems. *Math. Comp.*, 50:361–383, 1988.

[52] K. Eriksson and C. Johnson. Adaptive finite element methods for parabolic problems, I: A linear model problem. *SIAM J. Numer. Anal.*, 28:43–77, 1991.

[53] K. Eriksson and C. Johnson. Adaptive streamline diffusion finite element methods for stationary convection-diffusion problems. *Math. Comp.*, 60:167–188, 1993.

[54] K. Eriksson and C. Johnson. Adaptive finite element methods for parabolic problems, II: Optimal error estimates in $L_\infty L_2$ and $L_\infty L_\infty$. *SIAM J. Numer. Anal.*, 32:706–740, 1995.

[55] K. Eriksson and C. Johnson. Adaptive finite element methods for parabolic problems, IV: Nonlinear problems. *SIAM J. Numer. Anal.*, 32:1729–1749, 1995.

[56] K. Eriksson and C. Johnson. Adaptive finite element methods for parabolic problems, V: Long-time integration. *SIAM J. Numer. Anal.*, 32:1750–1763, 1995.

[57] K. Eriksson, C. Johnson, and S. Larsson. Adaptive finite element methods for parabolic problems, VI: Analytic semigroups. *SIAM J. Numer. Anal.*, 35:1315–1325, 1998.

[58] D. Estep. A posteriori error bounds and global error control for approximation of ordinary differential equations. *SIAM J. Numer. Anal.*, 32:1–48, 1995.

[59] D. Estep and D. French. Global error control for the continuous Galerkin finite element method for ordinary differential equations. *Modél. Math. Anal. Numér.*, 28:815–852, 1994.

[60] J. Frehse and R. Rannacher. Eine $L^1$-Fehlerabschätzung für diskrete Grundlösungen in der Methode der finiten Elemente. Tagungsband Finite Elemente, *Bonn. Math. Schr.*, 89:92–114, 1976.

[61] C. Führer. Error Control in Finite Element Methods for Hyperbolic Problems. Dissertation, Institute of Applied Mathematics, University of Heidelberg, 1996.

[62] C. Führer and G. Kanschat. A posteriori error control in radiative transfer. *Computing*, 58:317–334, 1997.

[63] C. Führer and R. Rannacher. An adaptive streamline-diffusion finite element method for hyperbolic conservation laws. *East–West J. Numer. Math.*, 5:145–162, 1997.

[64] M. B. Giles. On adjoint equations for error analysis and optimal grid adaptation. In Frontiers of Computational Fluid Dynamics 1998 (D.A. Caughey and M.M. Hafez, eds), pp. 155-170. , World Scientific, 1998.

[65] M. B. Giles and N. A. Pierce. Adjoint error correction for integral outputs. Technical Report NA-01/18, Oxford University Computing Laboratory, 2001.

[66] M. B. Giles, M. G. Larsson, J. M. Levenstam, and E. Süli. Adaptive error control for finite element approximations of the lift and drag coefficients in viscous flow. NA-97/06, Oxford University Computing Laboratory, 1997.

[67] P. Hansbo. Three lectures on error estimation and adaptivity. *Adaptive Finite Elements in Linear and Nonlinear Solid and Structural Mechanics* (E. Stein, ed.), Vol. 416 of CISM Courses and Lectures, Springer, 2002, to appear.

[68] P. Hansbo and C. Johnson. Adaptive streamline diffusion finite element methods for compressible flow using conservative variables. *Comput. Methods Appl. Mech. Eng.*, 87:267–280, 1991.

[69] R. Hartmann. A posteriori Fehlerschätzung und adaptive Schrittweiten- und Ortsgittersteuerung bei Galerkin-Verfahren für die Wärmeleitungsgleichung. Diploma thesis, Institute of Applied Mathematics, University of Heidelberg, 1998.

[70] R. Hartmann. Adaptive FE-methods for conservation equations. In Proc. *8th International Conference on Hyperbolic Problems. Theory, Numerics, Applications (HYP2000)* (H. Freistühler and G. Warnecke, eds.), pp. 495–503, Int. Series of Numer. Math. 141, Birkhäuser, Basel, 2001.

[71] R. Hartmann. Adaptive Finite Element Methods for the Compressible Euler Equations. Dissertation, Institute of Applied Mathematics, University of Heidelberg, 2002.

[72] R. Hartmann and P. Houston. Adaptive discontinuous Galerkin finite element methods for nonlinear hyperbolic conservation laws. Preprint 2001-20 (SFB 359), Universität Heidelberg, *SIAM J. Sci. Comput.*, to appear.

[73] R. Hartmann and P. Houston. Adaptive discontinuous Galerkin finite element methods for the compressible Euler equations. Preprint 2001-41 (SFB 359), Universität Heidelberg, *J. Comput. Physics*, to appear.

[74] F.-K. Hebeker and R. Rannacher. An adaptive finite element method for unsteady convection-dominated flows with stiff source terms. *SIAM J. Sci. Comput.*, 21:799–818, 1999.

[75] J. Heywood, R. Rannacher, and S. Turek. Artificial boundaries and flux and pressure conditions for the incompressible Navier-Stokes equations. *Int. J. Comput. Fluid Mech.*, 22:325–352, 1996.

[76] V. Heuveline and C. Bertsch. *On multigrid methods for the eigenvalue computation of non-selfadjoint elliptic operators.* East-West J. Numer. Math. 8, 275–297 (2000).

[77] V. Heuveline and R. Rannacher. A posteriori error control for finite element approximations of elliptic eigenvalue problems. Preprint 2001-08 (SFB 359), University of Heidelberg, *J. Comput. Math. Appl.*, 15:107–138, 2001.

[78] V. Heuveline and R. Rannacher. Adaptive finite element discretization of eigenvalue problems in hydrodynamic stability theory. Preprint, SFB 359, Universität Heidelberg, March 2001.

[79] P. Houston, R. Rannacher, and E. Süli. A posteriori error analysis for stabilized finite element approximation of transport problems. *Comput. Methods Appl. Mech. Eng.*, 190:1483–1508, 2000.

[80] T. J. R. Hughes and A. N. Brooks. Streamline upwind/Petrov Galerkin formulations for convection dominated flows with particular emphasis on the incompressible Navier–Stokes equation. *Comput. Methods Appl. Mech. Eng.*, 32:199–259, 1982.

[81] T. J. R Hughes, G. R. Feijoo, L. Mazzei, and J.-B. Quincy. The variational multiscale method – a paradigm for computational mechanics. *Comput. Methods Appl. Mech. Eng.*, 166:3–24, 1998.

[82] T. J. R. Hughes, L. P. Franca, and M. Balestra. A new finite element formulation for computational fluid dynamics, V: Circumvent the Babuška–Brezzi condition: A stable Petrov–Galerkin formulation for the Stokes problem accommodating equal order interpolation. *Comput. Methods Appl. Mech. Eng.*, 59:89–99, 1986.

[83] C. Johnson. *Numerical Solution of Partial Differential Equations by the Finite Element Method.* Cambridge University Press, Cambridge, 1987.

[84] C. Johnson. Adaptive finite element methods for diffusion and convection problems. *Comput. Methods Appl. Mech. Eng.*, 82:301–322, 1990.

[85] C. Johnson. Adaptive finite element methods for the obstacle problem. *Math. Models Meth. Appl. Sci.*, 2:483–487, 1992.

[86] C. Johnson. Discontinuous Galerkin finite element methods for second order hyperbolic problems. *Comput. Methods Appl. Mech. Eng.*, 107:117–129, 1993.

[87] C. Johnson. A new paradigm for adaptive finite element methods. In J. Whiteman, ed., *Proc. MAFELAP 93.* John Wiley, 1993.

[88] C. Johnson and P. Hansbo. Adaptive finite element methods in computational mechanics. *Comput. Methods Appl. Mech. Eng.*, 101:143–181, 1992.

[89] C. Johnson and P. Hansbo. Adaptive finite element methods for small strain elasto-plasticity. *Finite Inelastic Deformations - Theory and Applications* (D. Besdo and E. Stein, eds), pp. 273–288, Springer, Berlin, 1992.

[90] C. Johnson and R. Rannacher. On error control in CFD. Proc. Int. Workshop *Numerical Methods for the Navier-Stokes Equations* (F.-K. Hebeker *et al.*, eds), pp. 133–144, *vol. 47 of Notes Num. Fluid Mech*, Vieweg, Braunschweig, 1994.

[91] C. Johnson, R. Rannacher, and M. Boman. Numerics and hydrodynamic stability: Towards error control in CFD. *SIAM J. Numer. Anal.*, 32:1058–1079, 1995.

[92] C. Johnson and A. Szepessy. Adaptive finite element methods for conservation laws based on a posteriori error estimates. *Comm. Pure Appl. Math.*, 48:199–234, 1995.

[93] G. Kanschat. Parallel and Adaptive Galerkin Methods for Radiative Transfer Problems. Dissertation, Institute of Applied Mathematics, University of Heidelberg, 1996.

[94] G. Kanschat. Solution of multi-dimensional radiative transfer problems on parallel computers. *Parallel Solution of Partial Differential Equations* (P. Bjørstad and M. Luskin, eds), pp. 85–96, vol. 120 of *IMA Volumes in Mathematics and its Applications*, New York, 2000. Springer.

[95] G. Kanschat and R. Rannacher. Local error analysis of the interior penalty discontinuous Galerkin method. Preprint, Universität Heidelberg, 2002.

[96] H. Kapp. Adaptive Finite Element Methods for Optimization in Partial Differential Equations. Dissertation, Institute of Applied Mathematics, University of Heidelberg, 2000.

[97] G. Kunert. An a posteriori residual error estimator for the finite element method on anisotropic tetrahedral meshes. *Numer. Math.*, 86:471–490, 2000.

[98] G. Kunert. A posteriori $L_2$ error estimation on anisotropic tetrahedral finite element meshes. *IMA J. Numer. Anal.*, 21:503–523, 2001.

[99] G. Kunert. Edge residuals dominate a posteriori error estimates for linear finite element methods on anisotropic triangular and tetrahedral meshes. *Numer. Math.*, 86:283–303, 2000.

[100] P. Ladeveze and D. Leguillon. Error estimate procedure in the finite element method and applications. *SIAM J. Numer. Anal.*, 20:485–509, 1983.

[101] M. G. Larson. A posteriori and a priori error estimates for finite element approximations of selfadjoint eigenvalue problems. *SIAM J. Numer. Anal.*, 38:608–625, 2000.

[102] M. G. Larson and A. J. Niklasson. Adaptive multilevel finite element approximations of semilinear elliptic boundary value problems. *Numer. Math.*, 84:249–274, 1999.

[103] W. Liu and N. Yan. A posteriori error estimates for some model boundary control problems. *J. Comput. Appl- Math.*, 120:159–173, 2000.

[104] W. Liu and N. Yan. Local a posteriori error estimates for convex boundary control problems. Preprint, University of Kent, 2002.

[105] L. Machiels, A. T. Patera, and J. Peraire. Output bound approximation for partial differential equations; application to the incompressible Navier-Stokes equations. In S. Biringen, editor, *Industrial and Environmental Applications of Direct and Large Eddy Numerical Simulation*. Springer, Berlin Heidelberg New York, 1998.

[106] L. Machiels, J. Peraire, and A. T. Patera. A posteriori finite element output bounds for the incompressible Navier-Stokes equations: application to a natural convection problem. Technical Report 99-4, MIT FML, 1999.

[107] J. Nitsche. Ein Kriterium für die Quasi-Optimalität des Ritzschen Verfahrens. *Numer. Math.*, 11:346–348, 1968.

[108] C. Nystedt. A priori and a posteriori error estimates and adaptive finite element methods for a model eigenvalue problem. Technical Report Preprint NO 1995-05, Department of Mathematics, Chalmers University of Technology, 1995.

[109] J. T. Oden. Finite elements: an introduction. In *Handbook of Numerical Mathematics, Vol. II, Finite Element Methods (Part 1)* (P.G. Ciarlet and J.L. Lions, eds), pp. 3–15, North-Holland, Amsterdam. 1991.

[110] J. T. Oden and S. Prudhomme. On goal-oriented error estimation for elliptic problems: Application to the control of pointwise errors. *Comput. Methods Appl. Mech. Eng.*, 176:313–331, 1999.

[111] J. T. Oden and S. Prudhomme. Estimation of modeling error in computational mechanics. Preprint, TICAM, The University of Texas at Austin,
2002.

[112] J. T. Oden, W. Wu, and M. Ainsworth. An a posteriori error estimate
for finite element approximations of the Navier–Stokes equations. *Comput.
Methods Appl. Mech. Eng.*, 111:185–202, 1993.

[113] M. Paraschivoiu and A. T. Patera. Hierarchical duality approach to bounds
for the outputs of partial differential equations. *Comput. Methods Appl.
Mech. Eng.*, 158:389–407, 1998.

[114] R. Rannacher. Error control in finite element computations. *Proc. Summer
School Error Control and Adaptivity in Scientific Computing* (H. Bulgak and
C. Zenger, eds), pp. 247–278. Kluwer Academic Publishers, 1998.

[115] R. Rannacher. A posteriori error estimation in least-squares stabilized finite
element schemes. *Comput. Methods Appl. Mech. Eng.*, 166:99–114, 1998.

[116] R. Rannacher. *Finite element methods for the incompressible Navier-Stokes
equations.* Fundamental Directions in Mathematical Fluid Mechanics (G. P.
Galdi, J. Heywood, R. Rannacher, eds), pp. 191–293, Birkhäuser, Basel-
Boston-Berlin, 2000.

[117] R. Rannacher. Duality techniques for error estimation and mesh adaptation
in finite element methods. *Adaptive Finite Elements in Linear and Nonlinear
Solid and Structural Mechanics* (E. Stein, ed.), vol. 416 of CISM Courses and
Lectures, Springer, 2002.

[118] R. Rannacher and F.-T. Suttmeier. A feed-back approach to error control
in finite element methods: Application to linear elasticity. *Computational
Mechanics*, 19:434–446, 1997.

[119] R. Rannacher and F.-T. Suttmeier. A posteriori error control in finite element methods via duality techniques: Application to perfect plasticity. *Computational Mechanics*, 21:123–133, 1998.

[120] R. Rannacher and F.-T. Suttmeier. A posteriori error estimation and mesh
adaptation for finite element models in elasto-plasticity. *Comput. Methods
Appl. Mech. Eng.*, 176:333–361, 1999.

[121] R. Rannacher and F.-T. Suttmeier. Error estimation and adaptive mesh
design for FE models in elasto-plasticity. *Error-Controlled Adaptive FEMs
in Solid Mechanics* (E. Stein, ed.), John Wiley, to appear.

[122] S. Richling, E. Meinköhn, N. Kryzhevoi, and G. Kanschat. Radiative transfer
with finite elements I. Basic method and tests. *A&A*, 380:776–788, 2001.

[123] T. Richter. Funktionalorientierte Gitteroptimierung bei der Finite-Elemente-Approximation elliptischer Differentialgleichungen. Diploma thesis, Institute of Applied Mathematics, University of Heidelberg, 2001.

[124] M. Schäfer and S. Turek. Benchmark computations of laminar flow around a cylinder. (With support by F. Durst, E. Krause and R. Rannacher). *Flow Simulation with High-Performance Computers II (E. H. Hirschel, ed.), pp. 547-566, DFG priority research program results 1993-1995*, vol. 52 of Notes Numer. Fluid Mech., Vieweg, Wiesbaden, 1996.

[125] L. R. Scott and S. Zhang. Finite element interpolation of nonsmooth functions satisfying boundary conditions. *Math. Comp.*, 54:483–493, 1990.

[126] K. Siebert. An a posteriori error estimator for anisotropic refinement. *Numer. Math.*, 73:373–398, 1996.

[127] E. Stein and S. Ohnimus. Coupled model- and solution-adaptivity in the finite-element method. *Comput. Methods Appl. Mech. Eng.*, 150:327–350, 1997.

[128] F.-T. Suttmeier. Adaptive Finite Element Approximation of Problems in Elasto-Plasticity Theory. Dissertation, Institute of Applied Mathematics, University of Heidelberg, 1996.

[129] F.-T. Suttmeier. An adaptive displacement/pressure finite element scheme for treating incompressibility effects in elasto-plastic materials. *Numer. Meth. Part. Diff. Equ.*, 17:369–382, 2001.

[130] R. Verfürth. A posteriori error estimates for nonlinear problems. *Numerical Methods for the Navier–Stokes Equations (F.-K. Hebeker et al., eds), pp. 288-297*, vol. 47 of *Notes Numer. Fluid Mech.*, Vieweg, Braunschweig, 1993.

[131] R. Verfürth. A posteriori error estimates for nonlinear problems. Finite element discretization of elliptic equations. *Math. Comp.*, 62:445–475, 1994.

[132] R. Verfürth. *A Review of A Posteriori Error Estimation and Adaptive Mesh-Refinement Techniques*. Wiley/Teubner, New York Stuttgart, 1996.

[133] R. Verfürth. A posteriori error estimation techniques for nonlinear elliptic and parabolic pde's. *Rev. Eur. Élém. Finis*, 9:377–402, 2000.

[134] B. Vexler. A posteriori Fehlerschätzung und Gitteradaption bei Finite-Elemente-Approximationen nichtlinearer elliptischer Differentialgleichungen. Diploma thesis, Institute of Applied Mathematics, University of Heidelberg, 2000.

[135] C. Waguet. Adaptive Finite Element Computation of Chemical Flow Reactors. Dissertation, Institute of Applied Mathematics, University of Heidelberg, 2000.

[136] R. Zamni. Integrationsfehleranalyse bei der adaptiven Finite-Elemente-Methode. Diploma thesis, Institute of Applied Mathematics, University of Heidelberg, 2001.

[137] O. C. Zienkiewicz and J. Z. Zhu. A simple error estimator and adaptive procedure for practical engineering analysis. *Int. J. Numer. Meth. Eng.*, 24, 1987.

[138] M. Zlámal. On the finite element method. *Numer. Math.*, 12:394–409, 1968.

# Index

# Lectures in Mathematics  ETH Zürich

**LeFloch, P.G.** Hyperbolic Systems of Conservation Laws. The Theory of Classical and Nonclassical Shock Waves (2002) ISBN 3-7643-6687-7

**Valette, A.** Introduction to the Baum-Connes Conjecture (2002) ISBN 3-7643-6706-7

**Hélein, F.** Constant Mean Curvature Surfaces, Harmonic Maps and Integrable Systems (2001) ISBN 3-7643-6576-5

**Kreiss, H.-O / Ulmer Busenhart, H.** Time-dependent Partial Differential Equations and Their Numerical Solution (2001) ISBN 3-7643-6125-5

**Polterovich, L.** The Geometry of the Group of Symplectic Diffeomorphisms (2001) ISBN 3-7643-6432-7

**Turaev, V.** Introduction to Combinatorial Torsions (2001) ISBN 3-7643-6403-3

**Tian, G.** Canonical Metrics in Kähler Geometry (2000) ISBN 3-7643-6194-8

**Le Gall, J.-F.** Spatial Branching Processes, Random Snakes and Partial Differential Equations (1999) ISBN 3-7643-6126-3

**Jost, J.** Nonpositive Curvature. Geometric and Analytic Aspects (1997) ISBN 3-7643-5736-3

**Newman, Ch.M.** Topics in Disordered Systems (1997) ISBN 3-7643-5777-0

**Yor, M.** Some Aspects of Brownian Motion. Part II: Some Recent Martingale Problems (1997) ISBN 3-7643-5717-7

**Carlson, J.F.** Modules and Group Algebras (1996) ISBN 3-7643-5389-9

**Freidlin, M.** Markov Processes and Differential Equations: Asymptotic Problems (1996) ISBN 3-7643-5392-9

**Simon, L.** Theorems on Regularity and Singularity of Energy Minimizing Maps. based on lecture notes by Norbert Hungerbühler (1996) ISBN 3-7643-5397-X

**Holzapfel, R.P.** The Ball and Some Hilbert Problems (1995) ISBN 3-7643-2835-5

**Baumslag, G.** Topics in Combinatorial Group Theory (1993) ISBN 3-7643-2921-1

**Giaquinta, M.** Introduction to Regularity Theory for Nonlinear Elliptic Systems (1993) ISBN 3-7643-2879-7

**Nevanlinna, O.** Convergence of Iterations for Linear Equations (1993) ISBN 3-7643-2865-7

**LeVeque, R.J.** Numerical Methods for Conservation Laws (1992) Second Revised Edition. 6th printing 2002 ISBN 3-7643-2723-5

**Narasimhan, R.** Compact Riemann Surfaces (1992, 2nd printing 1996) ISBN 3-7643-2742-1

**Tromba, A.J.** Teichmüller Theory in Riemannan Geometry. Second Edition. Based on lecture notes by Jochen Denzler (1992) ISBN 3-7643-2735-9

**Bättig, D. / Knörrer, H.** Singularitäten (1991) ISBN 3-7643-2616-6

**Boor, C. de,** Splinefunktionen (1990) ISBN 3-7643-2514-3

**Monk, J.D.** Cardinal Functions on Boolan Algebras (1990)

## Department of Mathematics / Research Institute of Mathematics

Each year the Eidgenössische Technische Hochschule (ETH) at Zürich invites a selected group of mathematicians to give postgraduate seminars in various areas of pure and applied mathematics. These seminars are directed to an audience of many levels and backgrounds. Now some of the most successful lectures are being published for a wider audience through the Lectures in Mathematics, ETH Zürich series. Lively and informal in style, moderate in size and price, these books will appeal to professionals and students alike, bringing a quick understanding of some important areas of current research.

### http://www.birkhauser.ch

For orders originating from all over the world except USA and Canada:

**Birkhäuser Verlag AG**
c/o Springer GmbH & Co
Haberstrasse 7, D-69126 Heidelberg
Fax: ++49 / 6221 / 345 4229
e-mail: birkhauser@springer.de

For orders originating in the USA and Canada:

**Birkhäuser**
333 Meadowland Parkway
USA-Secaucus, NJ 07094-2491
Fax: +1 201 348 4505
e-mail: orders@birkhauser.com

*Birkhäuser*